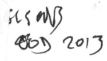

Joe:
Don't never forget!

J. O.

1993

The Patent Book

The Patent Book

AN ILLUSTRATED GUIDE AND HISTORY FOR INVENTORS, DESIGNERS AND DREAMERS

By James Gregory and Kevin Mulligan

A & W PUBLISHERS, INC. NEW YORK

Produced and prepared by Quarto Marketing Ltd.

Editor: Anne Ziff
Associate Editor: Nancy McNally
Production Manager: Tammy O'Bradovich
Typography: Gregory Enterprises

First published in the United States of America in 1979 by A & W Publishers, Inc.
95 Madison Avenue
New York, New York 10016
By arrangement with Quarto Marketing, Ltd.
Library of Congress Catalog Card Number: 78-68388
ISBN: 0-89479-037-4

Printed in the United States of America

To my mother.
                    J.G.

And mine, too.
                    K.M.

ACKNOWLEDGMENTS

Alison and Evelyn Clyburn for research assistance, Ellen Krieger for her help and editorial assistance, Franklin D. Wolffe for his assistance on the legal aspects of filing a patent.

CONTENTS

PART ONE

THE FAME AND FORTUNE SYNDROME

INTRODUCTION

Many people spend their lifetimes looking for an easy way to make a fast buck. There are always people laboring over some "brainchild" that they hope will be an overnight spectacular success, be it a "get-rich-quick" scheme or a drawer full of lottery tickets.

Among these hopefuls are the inventors, men and women who invest their time and talents in devising new or improved products that they hope will capture the imagination of the public and bring fame and fortune. This is not to suggest that greed and the desire to be famous are the fundamental motivation for creative and inventive thinking. They are, however, undeniable by-products of the inventive process and they provide the inventor with the needed impetus to keep experimenting when the going gets rough.

Belief in the marketability of his invention is what kept W.L. Judson at the drawing board trying to perfect his design for the zipper. The hope for riches certainly gave Eli Whitney the strength to rebuild when his cotton gin factory burned to the ground during his first year of business. Men such as George Westinghouse and Cyrus McCormick, Thomas Edison, and Alexander Graham Bell not only received the personal satisfaction of having advanced civilization through their inventions, but they also reaped tremendous financial rewards and their names and reputations are well known to us all.

Today, fame and fortune are just as accessible as they were one hundred years ago. There is a difference, however. Although the pitfalls remain the same, the rules of the game have changed somewhat. Be forewarned that it takes more—much more—than inventive genius to bring forth a successful, money-making brainchild. It takes common sense, a well-planned marketing strategy, manufacturing capabilities, advertising savvy, financial support, good contacts, and plain old luck.

The Central Telegraph Office in England, 1814.

C. Francis Jenkins was a pioneer in the development of motion pictures. The machine pictured above, called a radiovisor, was an early attempt at television.

The Inventor.

This is where *The Patent Book* comes in. The aim of this book is to give you a clear understanding of just how the patent system functions in America, to show you how to go about patenting your "brainchild," and to help you with your marketing strategy. In addition, *The Patent Book* provides the background to some of the patents that have changed the course of American and world history.

Nearly everyone has said to himself at one time or another: "I've got a great idea. I bet I could market it and make a million." No matter what your occupation—full-time inventor, designer, attorney, housewife, business executive, or student—*The Patent Book* is directed at the inventor you have lurking inside you.

"An entirely new system of television, gentlemen. It came to me in a dream."

THE HISTORY OF PATENTS

Although people have been inventing things for centuries, the idea of legally protecting an inventor's right of ownership is a relatively recent phenomenon. There was no patent office to record Prometheus's discovery of fire, or Icarus's birdlike wings, Leonardo da Vinci's myriad devices. Without a system of protective legislation, anyone could come along and steal an inventor's ideas and market them as their own.

The first known grant to an inventor awarded by a state was in 1421, in the Republic of Florence. Fifty years later, in 1474, an ordinance relating to patents was enacted in Venice. Other European countries followed suit. In the sixteenth century, the British monarchy doled out "monopolies" for all sorts of inventions, in addition to exclusive rights and privileges regarding the importation and establishment of new industries, and monopoly grants relating to known commodities. Ostensibly designed to protect an inventor's interest in his invention, this system was easily abused: Since no statute as such existed governing the monopolies, it was up to the whim of the ruler as to who should receive a grant. The recipients of these so-called monopolies were invariably friends or lovers of the grantor. The system served little useful function for the protection of an inventor's work; it was largely another example of political and social patronage.

Not until the seventeenth century was effective legislation enacted for the protection of both the invention and the inventor. Dissatisfaction over the arbitrary and elitist nature of the monopoly system caused Parliament to declare the practice unlawful. In 1624 a Statute of Monopolies was passed in the British Parliament which restricted the granting of monopolies for anything other than new inventions and confirmed the authority to grant exclusive rights for these inventions for a term of 14 years. By removing the power to grant monopolies from the monarchy and placing it in the hands of the British courts, this law successfully eliminated the abuses of the monopoly

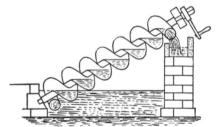

Archimedes' water-screw.

Early attempts to remove water from its source.

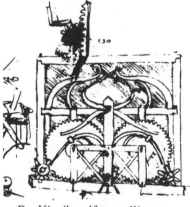

Da Vinci's self-propelling car.

The imagined Chinese land-sailing, four-wheeled wagons.

system, and served as the model upon which all subsequent patent legislation, both in Britain and in the United States, has been based.

The first patent bill in the U.S. was enacted by President George Washington on April 10, 1790. It gave the inventor the right to exclude all others from making, using, or selling his invention. This law was based on the British Statute of Monopolies which had been in existence in North America since the establishment of the first British colony. In 1641 an inventor named Samuel Winslow, of the colony of Massachusetts, was granted a "patent" by the Massachusetts General Court for his method of making salt. Five years later, the same court awarded the first "patent" on machinery to Joseph Jenkes, who invented a mill for the manufacturing of scythes. This procedure of court-awarded patents continued in the colonies for the next 140 years, until the delegates to the Constitutional Convention in Philadelphia in 1787, recognizing the importance of invention to economic development, adopted a provision whereby Congress was given the power "to promote the progress of science and useful arts by securing for limited times to authors and inventors the exclusive right to their respective writings and discoveries."

In January 1790, in an address to the second session of the First Congress, President Washington urged the representatives to give "effectual encouragement...to the exertion of skill and genius at home." A committee was established to formulate a bill for patents and for copyrights. The patent bill was presented for debate on February 16, and was signed into law on April 10. The Patent Act of 1790 defined the subject matter of a patent as "any useful art, manufacture, engine, machine, or device, or any improvement thereon not before known or used." In order to apply for a patent, the inventor had to present a specification and a drawing, and, if available, a model of the invention to the U.S. Patent Board.

The Patent Board, under the jurisdiction of the State Department, had the power to issue a patent "if they shall deem the invention or discovery sufficiently useful and important." The duration of the patent was not to exceed fourteen years. When Congress enacted the patent law in 1790, Thomas Jefferson was secretary of state, and as such he became the first administrator of the American patent system. He was joined on the Patent Board by Henry Knox, secretary of war, and Edmund Randolph, attorney general.

An early time keeping mechanism.

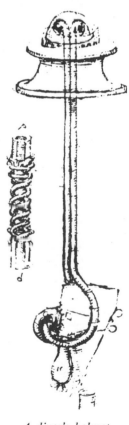

A diver's helmet.

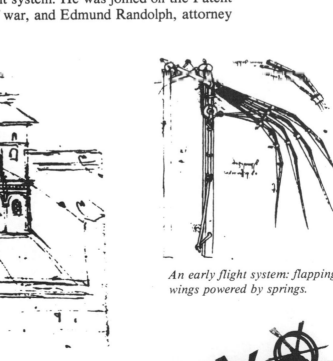

An early flight system: flapping wings powered by springs.

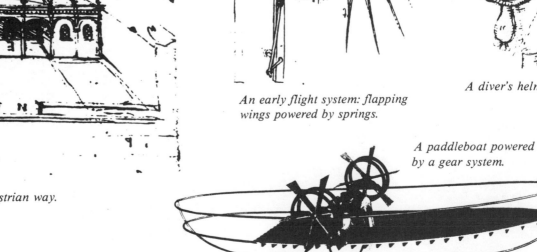

An elevated pedestrian way.

A paddleboat powered by a gear system.

Jefferson took an active interest in all of the applications addressed to the Patent Board. An inventor himself, he was convinced that "an inventor ought to be allowed a right to the benefit of his invention for some certain time. Nobody wishes more than I do that ingenuity should receive liberal encouragement." Because the patent law denied Patent Board members the right to apply for a patent, Jefferson never took out a patent for any of his inventions, but during his lifetime he invented several devices: a revolving chair, a folding chair or stool which doubled as a walking stick, a machine for treating hemp, and a pedometer, as well as an improvement on the moldboard of the plough which was significant in the agricultural development of the United States and for which he won a decoration from the French Institute.

On July 31, 1790, the first United States patent was granted to Samuel Hopkins of Pittsford, Vermont. Hopkins's invention was an improvement in "the making of Pot ash and Pearl ash by a new Apparatus and Process." Over the next three years, only fifty-five patents were awarded, due chiefly to the lack of time available to the board members for examining the inventions. In February 1793 the patent law was changed; no longer were applications judged for their importance and usefulness. Since time did not permit the Patent Board members to examine each invention in such detail, the board became a review board, but for the next forty-three years a patent was granted to anyone who applied for one, provided he submitted the proper drawings and paid the necessary fee. The function of the Patent Board became entirely clerical; since no one checked to see if a similar invention had been patented by someone else, all control over the protection of an inventor's rights was lost. It was during this period, on March 14, 1794 that Eli Whitney received his patent for the cotton gin.

On May 5, 1809, Mary Kies of Killingly, Windham County, Connecticut became the first woman to receive a patent. Her invention was a device for "weaving straw with silk or thread."

In 1836, on the basis of a comprehensive study of the patent office by Senator John Ruggles of Maine, the United States patent laws were revamped. The act of 1836 reestablished the "examination" system for granting patents. It established the Patent Office as a separate bureau within the Department of State, set the life of a patent at fourteen years with a possible seven year extension, and provided for a system of appeal, enabling an inventor who was denied a patent to argue for the acceptance of his application in the hopes of receiving a favorable decision.

Watt's first design for an oscillating engine.

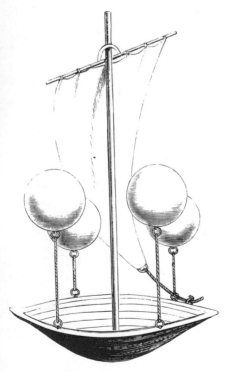

Jesuit Father Lana's Proposition.

Early telephone signals.

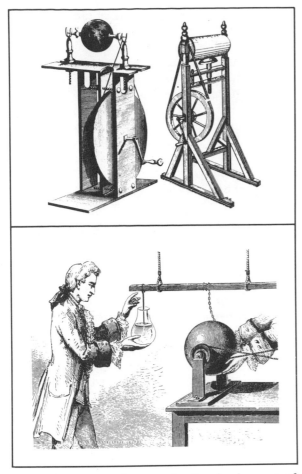

Electricity was an impetus for the development of many inventions in the 1700's. (Top) An Electrostatic machine; (bottom) a Hawkesbee machine to electrify water.

Thomas Jefferson, first administrator of the American patent system.

The United States Patent Office.

17

The Yankee Fleeing.

An amphibious tricycle.

Self-propelled water vehicle.

The act of 1836 serves as the basis for the present-day laws, although there have been several changes in the patent law since then. In 1842, an act permitted the patenting of designs for as long as seven years. Since then the law has been altered to permit the designer to request a duration of three and one-half, seven, or fourteen years.

In 1849, the Patent Office was transferred to the jurisdiction of the Department of the Interior, only to be transferred once again, in 1925, to the Department of Commerce, where it remains to this day. In 1859, copyright matters were incorporated into the Patent Office and in 1861, the length of a patent was changed to the present duration of seventeen years, with no extension.

A trademark law was enacted in 1881, but because it made no provision for the registration of trademarks, it was of little value. With the Trademark Act of February 20, 1905, registration of trademarks was first used in interstate commerce, thereby allowing companies to build up national reputations and maintain them on the basis of their good name.

A modification of the definition of what types of things could be patented was made in 1930 when the laws were extended to permit plants to be patented by anyone "who has invented or discovered and asexually reproduced any distinct and new variety of plant other than a tuber-propagated plant." This law was designed to help agriculture by stimulating the invention of new types of plants. The bill was championed by Thomas Edison who said, "nothing Congress could do to help farming would be of greater value and permanence than to give the plant breeder the same status as the mechanical and chemical inventors now have through the patent law."

Today the United States Patent Office employs an examining staff of over 1,200 people, who have been trained in different branches of science for the purposes of determining which inventions warrant patenting. This examining staff is divided into several Patent Examinimg Groups, each group handling one or more branches of industrial activity. It is the job of these examiners to research United States and foreign patents to determine if a similar patent has been issued. They also investigate scientific journals and publications to discover whether the idea under review has ever been mentioned. If the idea had been published, invented, or utilized previously, a patent cannot be issued.

Although the basic issuance fee for a patent is only $100, related costs, primarily court and legal fees, can run into the thousands of dollars. Furthermore, the days when one could "pay the fee and get the patent" are over. Out of 107,662 patent applications filed in one recent year, only 74,364 patents were granted. Due to the extensive examination involved in determining which ideas are patentable, the entire patent application procedure takes an average of two years to complete, and applying for a patent *does not* guarantee receiving one.

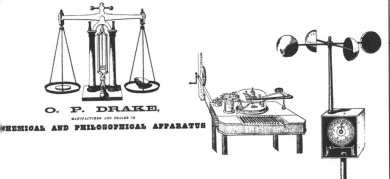

Advertisements for the marketing of new inventions in the 19th Century.

WHO NEEDS A PATENT?

You have come up with a unique and useful idea. You can make a working model of it and you think it has definite marketing possibilities. Now what happens?

The first thing to do is write a letter, called a disclosure document, to the Commissioner of Patents and Trademarks, Washington, D.C. 20231, detailing your invention and how it works. Once this letter along with the $10 fee is received by the Patent Office, it will be kept on file for two years, as evidence of when you actually thought of your new idea. Unless you make a public disclosure of your invention, you now have two years in which to file for a patent. Should you fail to follow up the letter by applying for a patent within those two years, the Patent Office will simply destroy the letter and you have lost nothing. Submitting the letter at the outset will serve as adequate proof of the date of your discovery and can be of great help should you become involved in a situation known as "interference."

Sometimes more than one inventor would file applications for the same invention at about the same time. Since the patent can be granted to only one of them, the Patent Office initiates proceedings to determine which inventor should receive the patent. These interference proceedings are concerned with two things: "conception of the invention" and "reduction to practice." Conception of the invention refers to the time when the inventor finally discovered how to make his invention work; reduction to practice refers to the actual construction of the invention in physical form. The inventor who proves to be the first both to conceive of the invention and reduce it to practice is considered the prior inventor.

Once you have submitted the disclosure letter, the next step is to search through the thousands of patents on file in the Search Room of the Patent and Trademark Office at Crystal Plaza, 2021 Jefferson Davis Highway, Arling-

"There you are – that's what they said they'd do: Railways, Road Services and Canals, all lumped together."

ton, Virginia to see if your great new idea has already been thought of by someone else. A patent cannot be granted on an invention that has been known or been used previously by others in the United States, that has been patented or described in a printed publication anywhere in the world, or that has been in public use or on sale in the United States more than one year before the date the applicant files for a patent. The only way to find this information is to search the files, and this can be an extremely time-consuming and expensive undertaking. In addition to the Search Room of the patent. The sample patent application at the end of this chapter will clarify this entire procedure. When listing more than one claim, you may place them in dependent form, in which each claim refers back to and further restricts a single preceding claim.

In the specification, you must include a description of the invention and of the manner and process of making and using it. The language must be such that any person skilled in the art to which the invention pertains to will be able to make it and use it. Be as specific as possible. The purpose of the description is to show the uniqueness of your invention, to distinguish it from other inventions similar to it and/or from the previous invention on which yours is an improvement.

Place the title of the invention, as briefly stated as possible, as a heading on the first page of the specification. Follow that immediately with a concise abstract of the technical disclosure of the specification in a separate paragraph entitled "Abstract of the Disclosure." Follow this with a short summary of the invention, indicating its nature and substance, perhaps stating the object of the invention, and proceed with the detailed description. You must conclude the specification with one or more claims that state exactly what you determine your invention to be and for what you are seeking the patent. When listing more than one claim, you may place them in dependent form, in which each claim refers back to and further restricts a single preceding claim.

The next part of the application procedure is the illustration or drawing, necessary only in those cases where a drawing is possible. The Patent Office has strict regulations regarding these drawings. They must show every feature of the invention specified in the claims. They must conform to a uniform style and they must be readily understood by persons using the patent descriptions. A detailed guide to the standards for drawings is available from the Patent and Trademark Office, Washington, D. C. 20231. One important note: if the inventor knows of any pertinent prior art dealing with his invention, he should call it to the attention of the Patent and Trademark Office in order that his patent will not subsequently be declared invalid on the basis of fraud.

The third and final part of the application procedure is the filing of the fee. The basic fee is $65. Should your description contain more than ten claims, each additional claim will cost you $2. Two or more independent claims per form will cost you an additional $10 per claim. The Patent Office defines a claim as dependent "if it incorporates by reference a single preceding claim which may be an independent or a dependent claim, and includes all the limitations of the claim incorporated by reference." Any claim not fitting this definition is considered to be independent.

At this point in your patent application procedure, it is advisable to enlist the services of a qualified patent attorney. There are more than 9,000 registered patent attorneys in the United States, and although they are prohibited by law from advertising for patent business, you can find a listing for them in your local Yellow Pages, and the Patent Office maintains a register of patent attorneys and patent agents. To be admitted to this register, a person must comply with the regulations prescribed by the Patent and Trademark Office, which now requires proof that the person is of good moral character and of good repute, in addition to having the legal and scientific/ technical qualifications necessary to enable him to provide valuable sevice to the applicant for a patent.

BEWARE OF THE INVENTION PROMOTION COMPANY. A growing number of organizations advertise their services for helping budding inventors to patent their inventions and then to market them. Let it be said right now that you neither need nor want their services. These companies are not registered with the Patent Office; they do not fulfill what they promise. Get yourself a good patent attorney and heed his advice.

Once you have engaged an attorney to conduct a patent search for you at a cost of anywhere from $100 to $300, and he has found no mention of another invention like your own, you are ready to proceed with your application for a patent. According to the patent law, only the inventor may apply for a patent. There are, however, certain exceptions. If the inventor is dead, his legal representatives may apply in his place; if he is insane, his legal guardians may apply for him; if he is alive but refuses to apply for a patent, then someone having a proprietary interest in the invention may apply on his behalf. In order to have your attorney handle the application procedure for you, it is necessary to sign a power of attorney, authorizing him to act legally in your stead.

The patent application consists of three parts, all of which must be on file with the Patent Office before the application will be accepted. The first thing you will need is a written document, called a specification, which describes your invention and outlines the claims for which you are seeking a patent. This must be accompanied by an oath or a written declaration in which you state that you believe yourself to be the original and first inventor of the subject matter of the application. This oath must be sworn to by the inventor before a notary public and the declaration must carry the full signature of the inventor plus the date.

When you have your application in order, mail it to the Commissioner of Patents and Trademarks, Washington, D. C. 20231, then sit back and wait. The fate of your patent is now in the hands of the Patent Office. The office processes the applications in the order in which they have been filed or in accordance with examining procedures established by the commissioner of patents. The Patent Office examiner studies the application and conducts his own in-depth search. All of this takes about two years to complete.

Once the examiner has reached his decision, the applicant is notified in writing of the action taken and the reason for this action. This notification is sent to the applicant's attorney or agent. Should your patent be denied, you have six months in which to request a reexamination. If this second examination results in another rejection, the office action is usually considered final and your recourses are (1) to make an appeal to the Board of Appeals, which consists of the patent and trademark commissioner, the assistant commissioners, and fifteen examiners-in-chief; (2) to appeal to the Court of Customs and Patent Appeals; or (3) to file a civil action against the patent commissioner in the United States District Court for the District of Columbia.

If, however, your application is accepted, you will be mailed a notice of allowance and will be required to pay the issuance fee within three months. The basic fee is $100; each additional page of specification costs $10, and each page of drawing costs $2. The awarding of your patent grants you "the right to exclude others from making, using or selling the invention throughout the United States," its territories and possessions for a term of seventeen years. The only way this term can be extended is by a special act of Congress.

Congratulations. You now have seventeen years in which to turn your newly-patented invention into an indispensable part of the American way of life. Remember, the patent gives you the right to exclude others from making, using, or selling your invention. It is not a divine right that will suddenly enable you to start mass-producing your product and selling it nationwide.In order to market your new product successfully, you need to have either a shrewd business sense and lots of money, or to be able to interest someone who does have these assets in manufacturing and marketing your product for you. Chapter III, "The Selling Strategy Behind the Invention," will give you some good ideas on how to accomplish this.

The letters patent for Alexander Graham Bell's improvement on the telephone.

THE SELLING STRATEGY BEHIND THE INVENTION

Once you have patented your invention, the next step is to make for yourself a clear and honest evaluation of your financial status and of the marketability of the invention. It is important to decide what expectations you, as the inventor, have for the invention you created. On the long road to success, this is a critical decision point; it is your last chance to get out of the project without investing a large amount of money. If you think it cost a lot to patent the invention, wait until you try to capitalize on it. Knowing what to expect from your efforts will help you evaluate the risks involved. This chapter spells out some methods for marketing your invention; it outlines what types of returns can be expected from an investment.

There are two ways to market your invention. If you have a lot of money and a shrewd business sense, you can manufacture and sell the product yourself. If not, then you must interest someone else in manufacturing and marketing the product for you. Since the majority of independent inventors possess neither a lot of money nor a shrewd business sense, let us discuss how to interest someone else in manufacturing and marketing your product.

It is not necessary to hire an outside sales consultant to create the material needed to sell your invention and it does not have to cost a lot to put together a presentation package for yourself. In preparing your own presentation, be certain to make the facts associated with the product clear and concise. A good format is suggested below:

Gregory Enterprises

Gentlemen:

I'd like to take this opportunity to acquaint you with our new development in structural design. The Bend-Two (trademark) system, a portable, inexpensive application of the groin vault, should be of great interest to manufacturers. It's flexible, strong and easily erected. Enclosed is more detailed information.

Gregory Enterprises offers an exclusive license to the manufacturers of specific applications, in return for a reasonable royalty against factory sales.

A copy of the Bend-Two patent, along with variations for which patents are now pending, will be forwarded at your request. Please feel free to call or write for answers to any questions you may have.

Sincerely,

James R. Gregory
President, Gregory Enterprises

Enclosures

<u>What we have</u> - A concept for a construction design which projects the
        classical groin vault from inexpensive, flat flexible materials.

<u>What it offers</u> - Vast applications as a portable, easily erected, strong
        and inexpensive structure.

<u>What you get</u> - An exclusive license for an application complementary to
        your product line.

<u>What it costs</u> - Our objective is long-term shared profit from successfully
        marketed products based on reasonable royalty against factory
        sales. We require a short term advance against royalties.

        The cost of building the system would vary according to the licen-
        see's needs and requirements, materials used, size of the structure,
        etc. (i.e., the specifications and cost of manufacturing a childs
        indoor playhouse would be completely different from those of a semi-
        permanent golf course shelter). Cost studies will be worked out
        according to those specific needs.

<u>SPECIFIC TERMS</u>
        -Royalty
        -Advance of $3,000 per month, to be credited against future royalties.
        -Approved work, travel, out-of-pocket expenses charged at cost.

<u>DESIGN, MANUFACTURING, AND ENGINEERING CONSIDERATIONS</u>
        Gregory Enterprises will provide a hand built prototype model designed
        to meet specific needs and requirements of the licensee at cost.

        Final design, and manufacturing up to licensee.

<u>PATENT PROTECTION</u>
        The Bend-Two$^{(TM)}$ system is patent pending.

        Confirmation obtainable:
            Fidelman, Wolffe, Leitner and Hiney
            Attorneys at Law
            Suite 300
            2120 L Street, N.W.
            Washington, D.C. 20037
            (202) 833-8801

Portable Groin Vault Applications

POTENTIAL USE AREAS

Playground

       Children's Tunnel Toy

       Playhouse (indoor)

       Playhouse (outdoor)

Sport Shelter

       Beach Cabin

       Camping Shelter

       Dressing Room & Bath House

       Ice Fishing Shelter

       Rain Shelter for Golf Course

       Swimming Pool Cabana

       Tennis Shelter (enclosed for year-round tennis)

       Tennis Shelter (rain and sun shelter)

       Warm-up Shelter for Ski Areas

       Warm-up Shelter for Ice Skating Rink (enclosed skating)

Chapel

       Gravesite Shelter

       Portable Chapel Outside Services

       Shelter for Catering Services

GREGORY ENTERPRISES

Portable Groin Vault Applications

POTENTIAL USE AREAS (continued)

Modular Building Units

       Bicycle Barn

       Bus Stop Shelters

       Concession Stands

       Garages

       Gas Station Canopies

       Greenhouses

       Hangars

       Homes

       Pavilions

       Portable Rest Rooms

       Sidewalk Shelter

       Storage Building

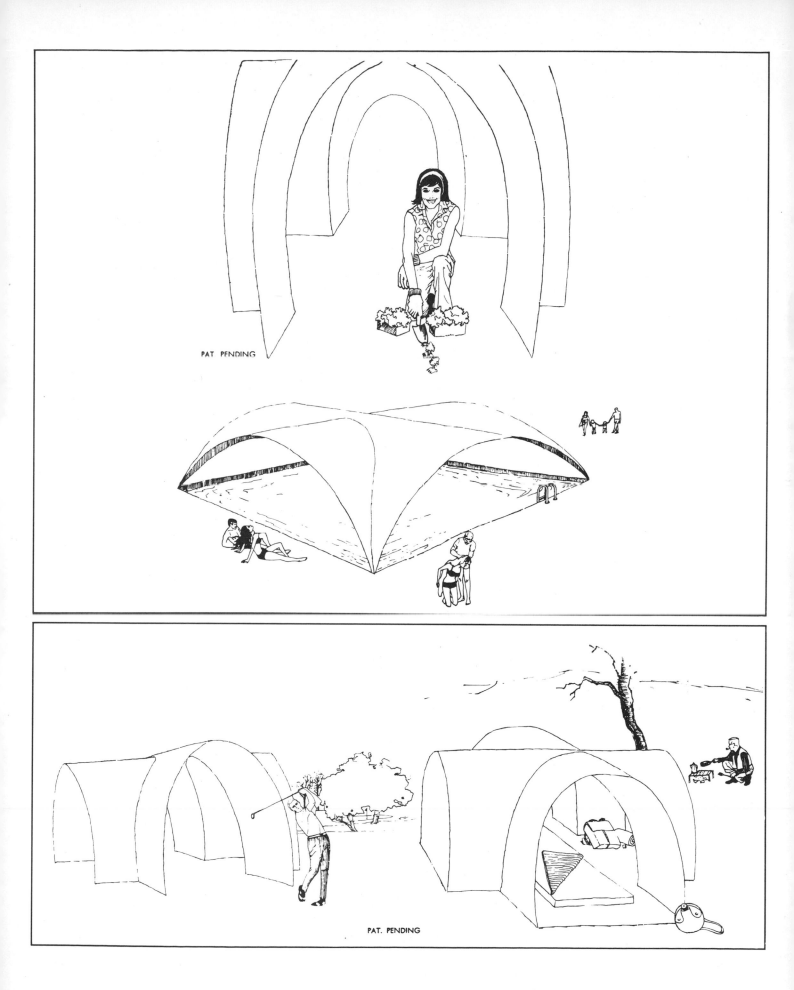

PAT PENDING

PAT. PENDING

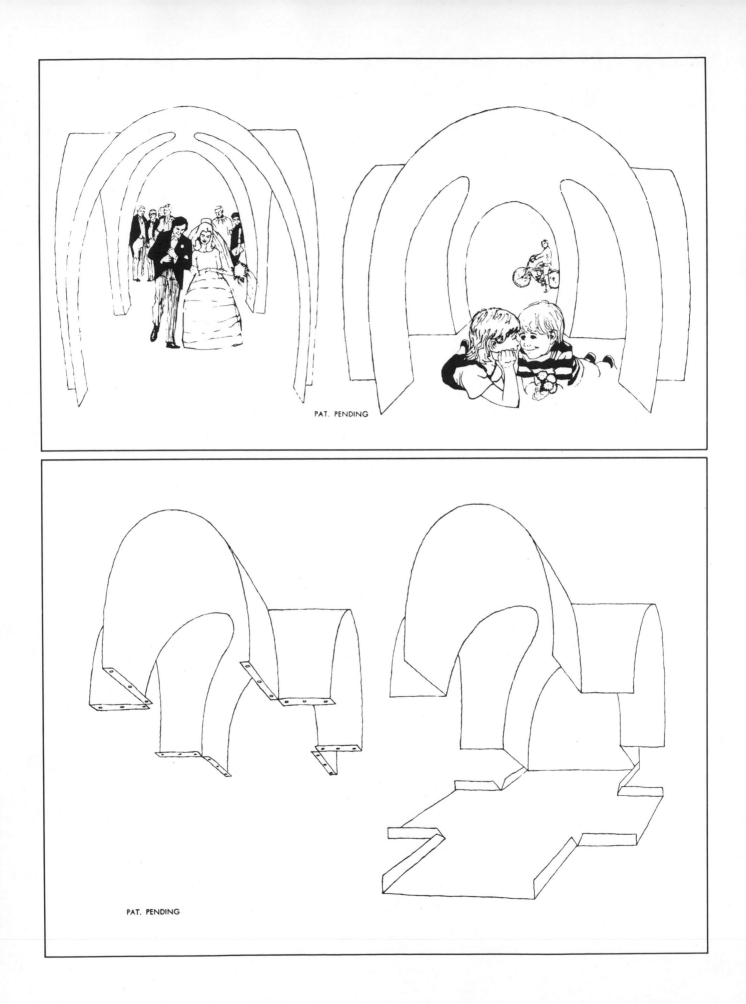

PAT. PENDING

PAT. PENDING

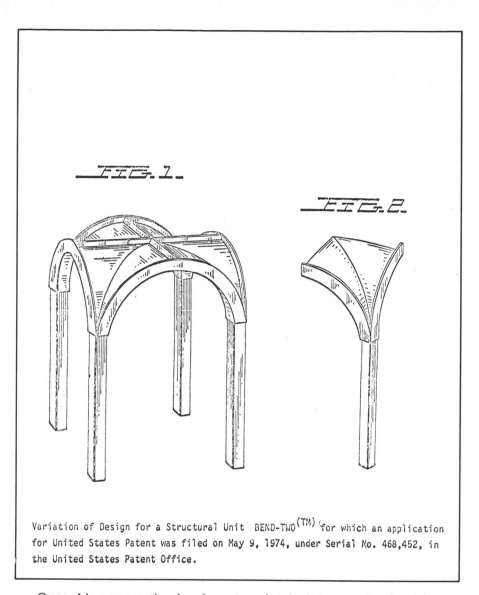

FIG. 1. **FIG. 2.**

Variation of Design for a Structural Unit BEND-TWO$^{(TM)}$ for which an application for United States Patent was filed on May 9, 1974, under Serial No. 468,452, in the United States Patent Office.

Once this presentation has been typed, take it to your local quick-copy center, where as many copies as you will need can be reproduced for a very reasonable price. You will not need more than 100 copies at first. Remember to include a self-addressed, stamped postcard with your presentation; it will usually insure a response.

When you have everything together and are ready to start sending your proposals to potential manufacturers, visit your local library and look up the *Thomas Register of Manufacturers.* This is an excellent book to assist you in your research of the companies that manufacture products similar to your own. These are the companies which are most likely to be interested in your new product.

Many large companies will ask you to sign a disclosure agreement before they will review your proposal. This absolves the company of any responsibility involving similar product designs which they may be developing. Most large companies have their own Research & Development departments; if they look at your invention before you sign a release and then discover that their own R & D people are working on a similar project at the same time, legal problems may result. It is up to the individual whether or not to sign a disclosure agreement. If you do not sign, you forfeit a chance to sell your good idea to that company; if you do sign, you risk the chance of that company utilizing your technology without paying for it.

"Our output has increased tenfold."

Be prepared to meet with the manufacturer to conclude a deal. Know what you want before you meet with him, and have your attorney present as the negotiations proceed. If the potential licensee has questions about the invention or its function, it is important that you be able to answer them quickly and fully in order to maintain his interest. It is a good idea to have a working model of the product with you when you meet with the manufacturer.

The ideal financial arrangement for the inventor is to receive money in advance (payable on signing the contract) and a royalty on sales. Royalties vary according to the volume and category of the product. Some royalties for low profit but high volume items are as low as one or two percent of the net selling price (salesman's commissions excluded). Royalties for higher priced but low volume products may be as high as ten percent. The average royalty is five percent.

It is wise to make yourself available as a consultant after the manfacturer has concluded an agreement. This may lead to additional income and acceptance of other inventions.

In some cases, the manufacturer may wish to purchase the rights of the invention without a long-term royalty arrangement. This should be discussed in detail with your lawyer. The man who designed the golden arches for Mc-Donald's sold his interest for a flat fee and no royalty. There is nothing worse than to see the product that you invented and sold for a few thousand dollars and know that, on a royalty basis, you would have made ten times that amount.

The temptation to manufacture and market the product by yourself is difficult to resist. Sometimes this seems like the easy way to avoid the maze of legalities and problems of licensing the product. Do not let the temptation deceive you into making a mistake that could cost you a life's savings. Manufacturing and marketing are highly professional fields which, in most cases, should be left to the professionals. The Small Business Administration has published a book by Robert W. James which provides detailed information about the steps required in the development of a new product and marks major decision points in that development.

Do not be afraid to quit for awhile if a project seems to be consuming your time and money. When you remove yourself from the problem for awhile, incubation can take place and a solution comes to mind almost "out of nowhere." Keep in mind that product development is a very expensive business. Proctor & Gamble is reported to have spent $70 million bringing their "Pringles's New Fangled Potato Chips" to market. And it is estimated that it will cost close to a quarter Billion dollars to develop something like the new Polaroid cameras. The most frustrating aspect of selling an idea is that success is so unpredictable. A product may have every logical potential for success and still fall flat. Other ideas may be totally absurd and become huge successes.

A word to the wise. We have seen a growing scandal brewing over the legitimacy of so-called invention brokers—companies that offer to sell your invention for you, for a "nominal fee" of course. That nominal fee averages between $1,000 and $1,500, and the services the brokers offer are, at best, suspect. One California broker, recently required to disclose its records of success, reported that out of more that "30,000 inventors it had contracts with, only three had earned a profit."

So now you want to know what the chances are of your success. The University of Oregon created an evaluation program for inventors that was carried out by faculty, students, and outside consultants. The results of the program showed that "from 100 initial ideas, fifteen may survive technical screening; eight may seem commercially feasable; five might then be developed; but probably only one would survive testing and commercialization to become a successful product."

No matter how you look at it, when you set out to market your invention, you are taking a gamble. Inventions we consider fundamental aspects of modern living, such as the zipper and the typewriter, were incredibly difficult to market when they were first invented. Men like Whitcomb L. Judson sank fortunes into their inventions and died broke and defeated. Other inventors, like Henry Ford and George Westinghouse, made fortunes from their inventions and lived to enjoy both fame and wealth. The next section, which presents an illustratred view of some fifty inventions and their inventors, describes some of the obstacles and rewards the present-day inventor in America can expect to encounter once he has made the decision to go ahead and attempt to capitalize on an invention.

A domestic contrivance, invented by Smith, and available — free of charge — to all readers of the Agriculturalist.

A proposed submarine railroad boat.

PART TWO

FAMOUS PATENTS

ELI WHITNEY

COTTON GIN

Not every inventor has difficulty acquiring a patent. All Eli Whitney had to do was submit drawings of his cotton gin to Thomas Jefferson with a sworn affidavit as to their originality. In addition to being secretary of state in 1794, Jefferson was in charge of the Patent Office.

Jefferson requested that Whitney bring a working model of his cotton gin to Philadelphia, which Whitney did, and the patent was issued by President Washington on March 14, 1794. What proved difficult for Whitney and his partner, Phineas Miller, was enforcing the patent, an activity which kept them involved in lawsuits for over ten years.

Whitney was born in Westborough, Massachusetts, on December 8, 1765. His mother died when he was twelve. His father remarried but Whitney and his stepmother did not get along. Whitney's father, however, was very encouraging of young Eli's talents and extremely supportive of his interest in tools and machinery. Whitney spent his childhood working on the family farm and in his own workshop. He was not educationally

ambitious, but early on he realized that college offered a way out of the provincial life of Westborough and would help him to acquire a certain amount of financial success. After four years working as a schoolteacher to finance his college education, Whitney enrolled at Yale in the spring of 1789. His father, against the objections of his stepmother, provided $1,000 towards his son's tuition.

While at Yale, Whitney studied grammar, composition, debating, and theology, but his courses in law, mathematics, and science proved of much greater interest to him. He persevered in all his studies, however, and at the end of four years graduated with a Bachelor of Arts degree, and a teaching job offer which the president of Yale, Ezra Stiles, secured for him.

Whitney traveled south in the company of Mrs. Catherine Greene, the widow of General Nathanael Greene, and Phineas Miller, a Yale graduate and former tutor of the Greene children, who supervised Mulberry Grove, the Greene plantation outside Savannah, Georgia. When Whitney and his pro-

Whitney's original cotton gin, patented March 14, 1794.

spective employer, a Major Dupont of South Carolina, could not agree on a salary, Whitney ended up staying on at Mulberry Grove with Miller and Mrs. Greene for several months. It was during this time that he became aware of the need to modernize the primitive procedures for picking the seeds out of the cotton fiber. On the average, a slave could manage to pick seeds out of approximately one pound of short-staple cotton fiber per day. Clearly, this was not an economically viable way to run the South's most important and potentially lucrative industry.

So Whitney went to work. He constructed a machine with slots wide enough to let the cotton fiber through, but narrow enough to prevent the seeds from passing through as well. The basic construction was that of a wooden box with a metal hand-cranked cylinder bearing several hooks. As the cylinder was turned, the hooks caught the cotton fibers and pulled them through the slots, while the seeds fell past to the bottom of the box. Once the seeds had been removed, stiff brushes, which rotated with the continued turning of the cylinder, separated the cotton fibers from the hooks. Using Whitney's original prototype, a slave could increase his production from one pound a day to close to fifty pounds.

Excited by the financial possibilities of the new device, Whitney and Miller formed a partnership. Miller, backed by Mrs. Greene, put up the money to finance the marketing of Whitney's invention. Miller and Whitney split the profits. In June 1793, Whitney went north, to New Haven, Connecticut, to patent the cotton gin, which he did in the spring of the following year. That's when the troubles began.

It took time for Whitney to train workers to construct the machines; an epidemic in New Haven halted production

ELI WHITNEY

Using Whitney's original cotton gin prototype, a slave could increase his production from one pound a day to close to fifty pounds.

for a time; a year later his workshop burned to the ground, and he was forced to rely exclusively on Catherine Greene's continued and generous financial support. Unfortunately, the South couldn't wait for Whitney to get his business off the ground. Pirated models of the Whitney gin were built and sold; modifications were made. Complete cotton gin operations were established throughout Georgia and Phineas Miller spent most of his time battling unsuccessfully in endless lawsuits to protect his and Eli's patent. Finally, in 1801, the South Carolina legislature purchased the right to use one of Whitney's machines in the state ginneries. A similar arrangement followed a year later with North Carolina. But it wasn't until 1806, when Judge William Johnson of Georgia forbade the production of a cotton gin, which he determined to be an imitation of the Whitney model, that the entire legal controversy came to an end.

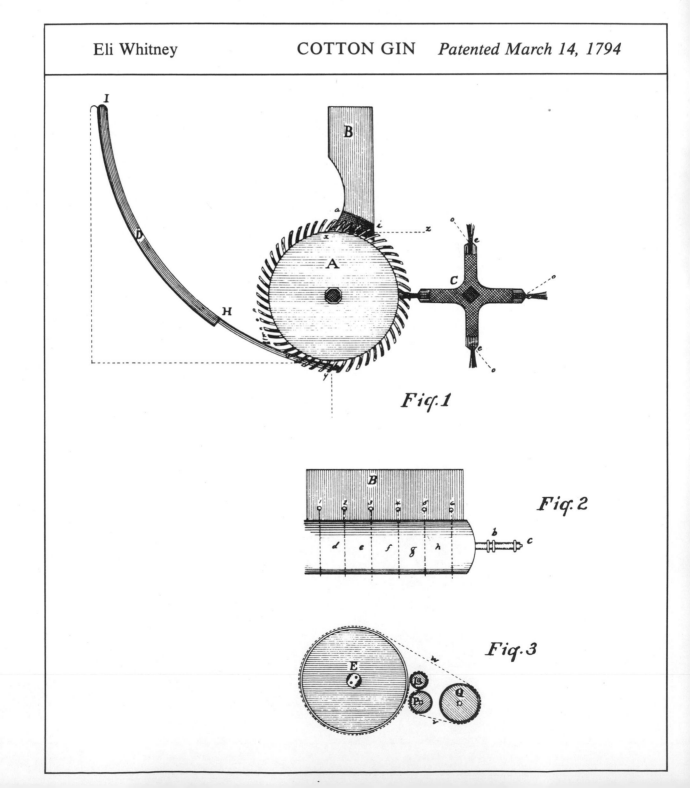

Eli Whitney   COTTON GIN   *Patented March 14, 1794*

Fig. 1

Fig. 2

Fig. 3

Illustration from Eli Whitney's patent application for a cotton gin.

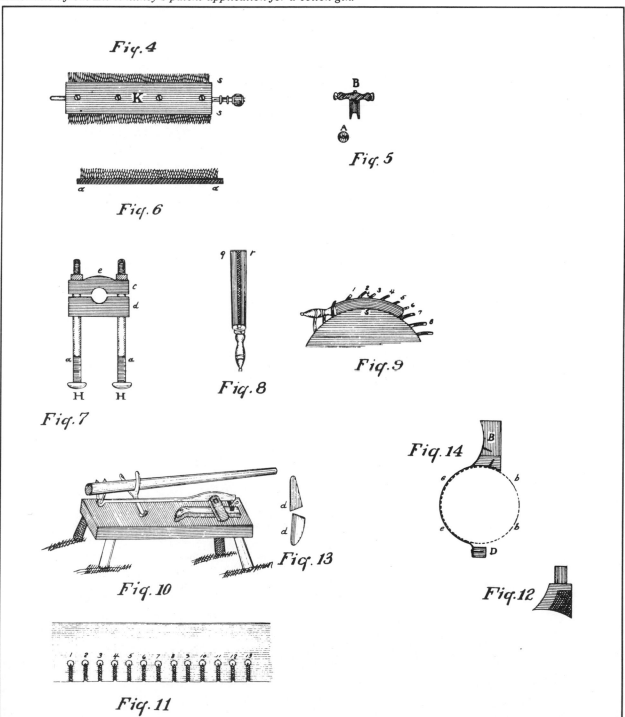

WILLIAM AUSTIN BURT
CHARLES LATHAM SHOLES

TYPEWRITER

For centuries people had been trying to invent a workable type writing machine. In 1714, Queen Anne of England issued a royal letters patent to one Henry Mill, an engineer, for his invention of a device described as "An Artificial Machine or Method for the impressing or transcribing of Letters singly or progressively one after another, as in Writing, whereby all writings whatsoever may be engrossed in Paper or Parchment so neat and exact as not to be distinguished from Print." Unfortunately no evidence exists of what Mill's machine looked like; there are no drawings, no model, nothing other than the patent record.

In the century that followed, several attempts were made to design a mechanical writing machine. Most of them were intended for the blind. In 1829, Louis Braille invented a machine that embossed letters and symbols to be read with the fingertips. And in that same year, the first American typewriter was invented.

William Austin Burt, a surveyor from Massachusetts, who fought in the War of 1812 and then moved with his family to the backwoods of Michigan, devised a machine to aid him in his work as a territorial legislator. Burt was becoming bogged down with the bureaucratic paperwork of a government employee. He went to his friend and local newspaper editor, John P. Sheldon, to borrow some type from the newspaper's composing room. He arranged the metal type in a semicircle within a large wooden box. By rotating the semicircle, the desired letter could be positioned at the exact spot on the paper for printing. A lever pressed the letter against the paper. Burt used pads for inking the type, and the paper was fed into it from one continuous roll which could be cut to page length. The entire machine was similar to the simple typewriter a child would play with today. Although Burt's typographer, as he called it, printed neatly and clearly, it was much faster to write longhand than to use the machine.

Burt's friend Sheldon, however, was wildly enthusiastic about the invention and, using the typographer, wrote a letter to President Andrew Jackson:

SIR:
This is a specimen of the printing done by me on Mr. Burt's typographer. You will observe some inaccuracies in the situation of the letters: these are owing to the imperfections of the machine, it having been made in the woods of Michigan where no proper tools could be obtained by the inventor ....I am satisfied...that the typographer will be ranked with the most novel, useful and pleasing inventions of this age.

Burt typed his own message to President Jackson on the back of the letter — a formal request for a patent. Burt received his patent, signed by both President Jackson and Martin Van Buren, on July 23, 1829.

It was now time to market the invention. Unfortunately, no one in the great northwoods was interested. Sheldon and Burt traveled south to New York City, looking for a backer. Again, no one was interested. So the two men traveled back to Michigan, Sheldon to his *Michigan Gazette*, Burt to his log cabin in the woods to work on yet another invention: a surveying compass that used the sun instead of the earth's magnetic attraction to find direction. What happened to that invention? It became United States government standard equipment for the next seventy-five years. And what happened to the typographer/typewriter? It underwent fifty more years of trial and error, until E. Remington & Sons of New York, the sewing machine manufacturers, came out with the "Sholes & Glidden Type-Writer." But that is another story altogether.

Alfred Ely Beach invented a typewriting machine that printed embossed letters and was used by the blind. He exhibited his invention at the American Institute Fair in 1856, and won a gold medal for his achievement.

William Austin Burt's original typewriter, which he patented on July 23, 1829.

In 1860, Christopher Latham Sholes was collector of customs in Milwaukee, Wisconsin. Married, with ten children and one grandchild, Sholes spent most of his time at a machine shop run by a man named C.F. Kleinsteuber. The shop was a clubhouse of sorts for a small group of men who fancied themselves amateur inventors. The men spent most of their time talking, but in between conversations, each of them worked on his own special project. Sholes' scheme was to devise a machine that automatically numbered the pages of books. One day, he read an article in *Scientific American* about the invention of a tero-type machine, which was similar to but more sophisticated than Burt's typographer that had been patented both in England and the United States in the 1860s. Sholes' associates urged him to try to come up with a similar device — but one that could also number book pages.

With the help of his friends, Carlos Glidden, S.W. Soule, Henry W. Roby, and Matthias Schwalbach, Sholes set to work. By the end of the summer he had come up with a working model of a typewriter, built into the top of an old table. He placed each letter on a separate bar or key instead of on a circular plate as Burt had done. The letters were imprinted on the paper by striking upward through a piece of carbon paper. Pleased, but dissatisfied, Sholes dismantled the machine to make improvements on it. He worked on these for several years.

All these alterations required money, and money was one thing the Kleinsteuber group did not have very much of. In order to finance further experimentation, they sold shares in the invention. An old newspaper acquaintance, James Densmore, from Pennsylvania, offered to pay the group's $600 in outstanding bills plus all future expenses for a twenty-five percent interest in the machine. They agreed.

In June 1868, the first Sholes-built typewriter was patented, but Densmore was not satisfied. Insisting that the machine could not possibly be marketed successfully unless major improvements were made, he coerced Sholes into making repeated changes. Within the next six years, that is until 1873, Sholes constructed and Densmore rejected some forty or so machines.

In the fall of 1870, Densmore and Sholes journeyed to New York in an attempt to sell the invention to the Automatic Telegraph Company for $50,000. Densmore arrived from Pennsylvania with powers of attorney from Glidden and Soule. Sholes came from Milwaukee with the typewriter. After several weeks of waiting, Densmore and Sholes were told no. Automatic Telegraph Company thought that one of their telegraph experts, Thomas A. Edison, could devise a much better machine for the same amount of money. In fact, what Edison came up with for their $50,000 was not a typewriter but an electrically operated machine that later turned into the stock ticker.

Densmore and Sholes returned home. By this time Densmore had invested every last penny he could muster in the invention and was forced to live like a pauper in cheap hotel rooms. Yet his belief in the typewriter never faltered. By 1872 he owned almost the entire interest in the typewriter. Glidden, Roby, Schwalbach, and most of the shareholders had turned their stocks over to him. The company was now basically Sholes and Densmore. The two men rented a studio outside Milwaukee where they manufactured much-improved machines. *Scientific American* assessed them and they sold rather well. Unfortunately, they cost more to manufacture than they could be sold for.

Densmore interested a promoter friend of his, George Washington Napoleon Yost, in taking a look at the machine and suggesting possible ways of marketing it successfully. Yost suggested selling the invention to E. Remington & Sons, the rifle, pistol, and sewing machine company.

The company loved it and, within a year, the first Remington-built typewriter was placed on the market. Encased in black enamel with gold swirls and little flowers emblazoned all over it, and with a treadle for returning the carriage, the machine resembled a sewing machine more than a typewriter. Stenciled on the side in gold letters was the machine's name: "Sholes & Glidden Type-Writer." What began as idle conversation at the Kleinsteuber machine shop had become a reality and the life of the typical office worker was about to undergo a drastic change.

A further refinement of the Remington-produced Sholes & Glidden typewriter.

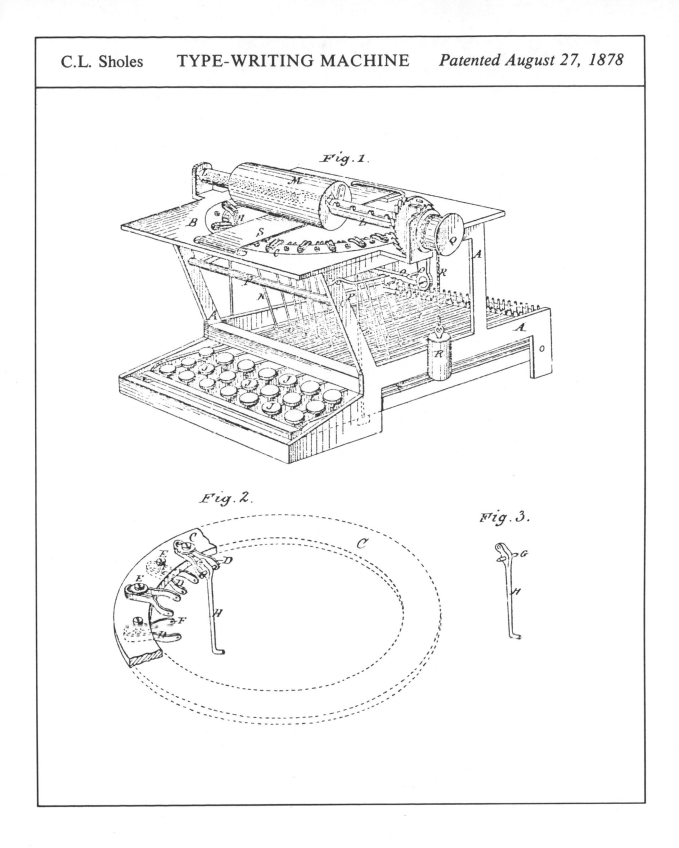

Fig.1.

Fig.2.

Fig.3.

CYRUS H. McCORMICK

REAPER

Cyrus McCormick was no stranger to the world of farming and farm machinery. His father, Robert McCormick, was an inventor of farm machinery and had played an important part in the early development of the reaper in the United States. Cyrus grew up assisting his father and watching him design and construct several farm implements, among them a hemp rake and a threshing machine. When, after twenty years of work, his father gave up on his efforts to make a machine that would reap grain mechanically, Cyrus constructed his own crude reaper, based on entirely different principles from his father's, and used it successfully on his own farm. He continued to make improvements on this reaper, and, in 1832, at the age of 23, exhibited his invention on several farms in the area.

At the same time, a man named Obed Hussey constructed a successful reaping machine in Maine. McCormick quickly applied for a patent, which he received on June 21, 1834. From 1837 to 1845, McCormick's father manufactured his son's invention on a contract basis, along with several other licensees. But McCormick had difficulty with some of the licensees, and upon his father's death in 1846, he settled in Chicago, where he established his own factory. By 1850, he had developed a nationwide business. He was forced to compete with Hussey's reaper as well as about thirty other rivals; nevertheless, by constantly making improvements on his machine and employing excellent business methods, he was able to fend off his competitors and expand his company to such an extent that he soon brought his two brothers, Leander J. and William S., into the firm to manage it.

McCormick was definitely an inventor who knew how to market his product. He was among the first salesmen to initiate such revolutionary business practices as guarantees, field trials, and deferred payments for merchandise. Through his efforts, the McCormick Harvesting Company grew from a small family business to a worldwide corporation.

C. H. McCormick   **PLOW**   *Patented November 19, 1833*

In 1858 he married a woman named Nettie Fowler, who proved to be as astute in business as he was. Throughout their life together, she remained his constant consultant, and upon his death in 1884 she assumed the leadership of his business — the first American woman to head a major corporation. In 1902 the McCormick Harvesting Company joined with several other companies, among which was Deering Harvester, to form International Harvester. Cyrus McCormick, Jr. became the company's first president.

CYRUS H. McCORMICK

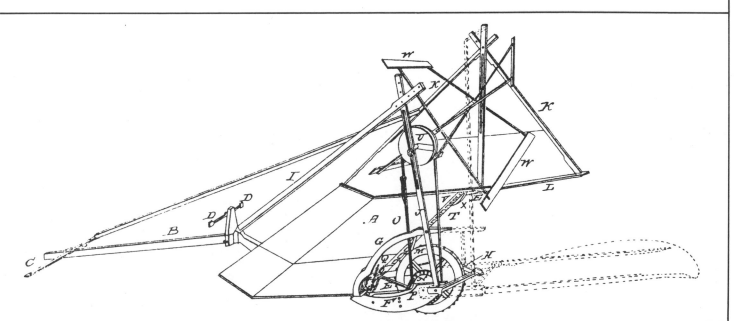

| C. H. McCormick | **REAPER** | *Patented June 21, 1834* |

Although his father had first experimented with designs for a machine that would reap grain mechanically, Cyrus McCormick was the first person to construct a reaper that worked. An acute business sense helped McCormick develop his idea into a world-wide business.

AMUEL COLT

R E V O L V I N G   G U N

Samuel Colt would have missed out on a fortune but for the Mexican War. In the 1830s, on a voyage from Boston, Massachusetts, to India, Colt started whittling a model of a revolver out of wood. When he returned to the states a year later, he fashioned similar models out of metals.

Between 1831 and 1835, Colt worked on many different versions of the metal models and applied for patents for several of them. In 1836 he received his first United States patent for a single-barreled pistol / rifle which had a multichambered breech that rotated and could be turned, locked, and unlocked by cocking. He had taken out English and European patents for the same device a year earlier.

In March 1836 Colt formed the Patent Arms Manufacturing Company in Paterson, New Jersey. Unfortunately, he had an impossible time trying to generate business. By 1842 his business had failed, and he had lost control of his patents.

Colt went into the telegraph business and devoted his spare time to the invention of electrically discharged submarine mines. With the outbreak of war in Mexico, his fortune changed. He re-ceived an order from the United States government for 1,000 pistols and quickly regained possession of his patent. He immediately started the manufacture of firearms.

These first guns were made at Eli Whitney's factory in Whitneyville, Connecticut. Towards the end of 1847, Colt was able to set up his own munitions factory in Hartford, Connecticut. Business prospered. In 1855 he established the largest private armory in the world on the site of the present Colt plant alongside Interstate 91 in Hartford. Atop the factory in Hartford is a Turkish cupola, given to Colt by a sultan of Turkey in gratitude for the arms sold to him during a provincial war.

Colt, who had been born in Hartford and then run off to sea at the age of sixteen, amassed one of the largest fortunes of his time, largely through innovative business and production techniques. He was one of the first manufacturers successfully to use the system of interchangeable parts and the production line to reduce costs and step up productivity.

When he died in Hartford in January 1862 at the age of 48, Colt was recognized as an independent and progressive

The Romans invented a weapon called a ballista that hurled stone cannon balls at the enemy's ranks.

SAMUEL COLT

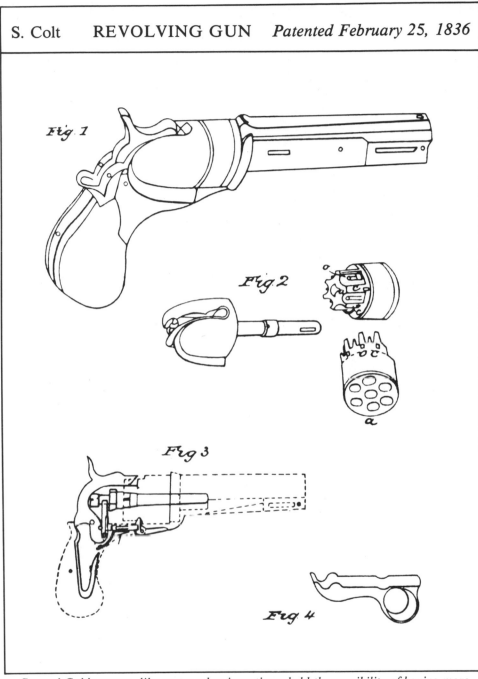

Samuel Colt's patent, like many other inventions, held the possibility of having more than one application. On these pages, Colt's revolving gun is shown being used both as a pistol and as a rifle. All three of the pages appeared in the original patent, illustrating Colt's foresight; he was well aware of the wide variety of applications his invention offered.

thinker, having been one of the first company owners to be concerned with company working conditions and employee welfare. But if it hadn't been for the advent of the Mexican War and that fortuitous order for 1,000 pistols, the revenue and profit from Colt's patents might have gone to someone else.

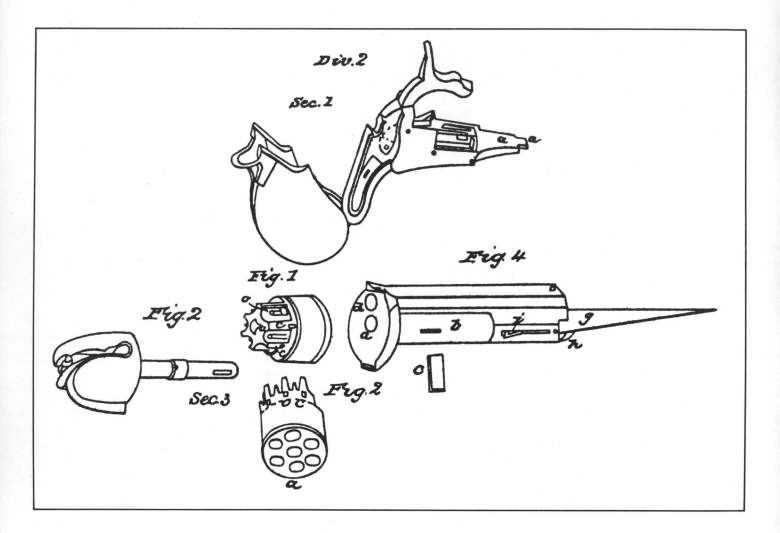

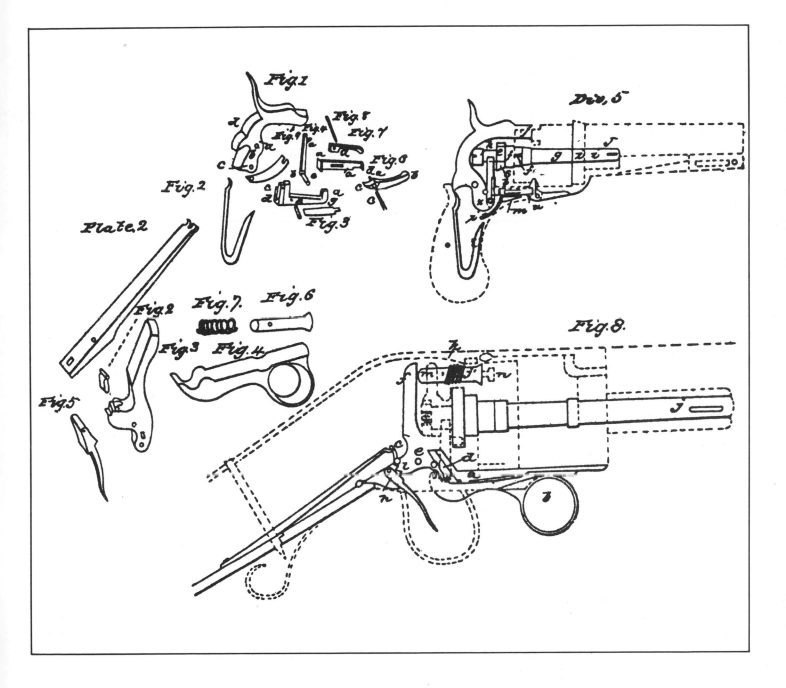

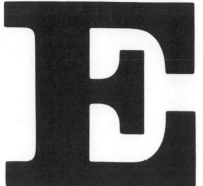

ELIAS HOWE JR.
ISAAC SINGER

SEWING MACHINE

Elias Howe was working as a machinist in a shop in Boston when he overheard a conversation that gave him the idea of inventing a machine that could sew. He devoted all his time to the device.

Unfortunately, Howe had very little money and his health was poor. He was forced to board with a friend, who agreed to support Howe's family in exchange for a half interest in the invention. In order to bring in more money, Howe's wife took in hand-sewing. After years of trying to duplicate the action of the human hand in sewing, Howe finally took a different tack and in May 1845 came up with a workable model of the sewing machine. This first machine employed an eye-pointed needle that made loops of thread through which the thread loaded on a shuttle underneath the fabric passed, making a lockstitch.

The first thing Howe did, upon securing his patent on September 16, 1846, was to make two suits, one for himself and one for a friend, in order to show exactly how strong the stitching was. Unhappily for him, no one in America seemed the least bit interested in this invention. In fact, many people were angered by it, afraid that it would take jobs away from the hand-sewing industry. Howe's brother was therefore commissioned to market the machine in England. He succeeded in interesting a London manufacturer of corsets and umbrellas in it and he leased the English rights for £250. The English company, which also manufactured leather bags, offered Howe a job at £3 per week to adapt the machine to stitch leather, and in 1847 Howe and his family sailed to England. But before long Howe quarreled with his employer and quit, where-

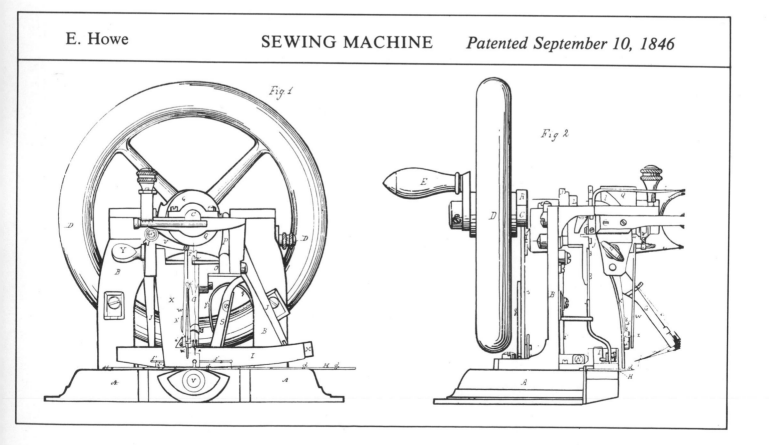

E. Howe   SEWING MACHINE   *Patented September 10, 1846*

Fig 1

Fig 2

upon he sent his family back to the States. He made his own way back by hiring himself out as cook on an emigrant ship. Howe arrived home to find his wife seriously ill; to make matters worse, all their possessions had been lost at sea.

His wife died within the year. Penniless, Howe looked about and saw that in the two years since he had left New York, a tremendous interest in a mechanical sewing machine had developed. He also discovered that there were several pirate editions of his machine being marketed. With his partner, George W. Bliss, he now waged a long and successful campaign to uphold his patent, and in 1834 the courts awarded him a royalty on all sewing machines sold. From abject poverty in 1849, Howe quickly emerged as a multimillionaire. Before the expiration of his patent, it is estimated that he received more than $2 million in royalties.

During the Civil War, Howe fought as an enlisted man with a Connecticut regiment. On one occasion, when the company pay was delayed, he advanced the money himself. He died in 1867 at the age of 48 in Brooklyn, New York.

Although Elias Howe is credited with the invention of the sewing machine, the man who made it a household item was Isaac Singer. Motivated by money ("I don't give a damn for the invention," he is quoted as saying, "the dimes are what I'm after") Singer was an excellent salesman.

Taking the sewing machines that were on the market in 1850, he made several improvements of his own and came out with the first machine that really worked well. He then began a worldwide publicity campaign. One of the major stumbling blocks at the time was convincing the public that women were *not* too stupid to operate the sewing machine. Singer employed beautiful women to sew on Singer sewing machines in store windows; he dispatched salesmen throughout the country to knock on doors. He and his partner established a worldwide network of repair shops. Contrary to what at that time were considered sober business practices, Singer sold his machines on an installment plan. In a short while, he had

ELIAS HOWE, JR.

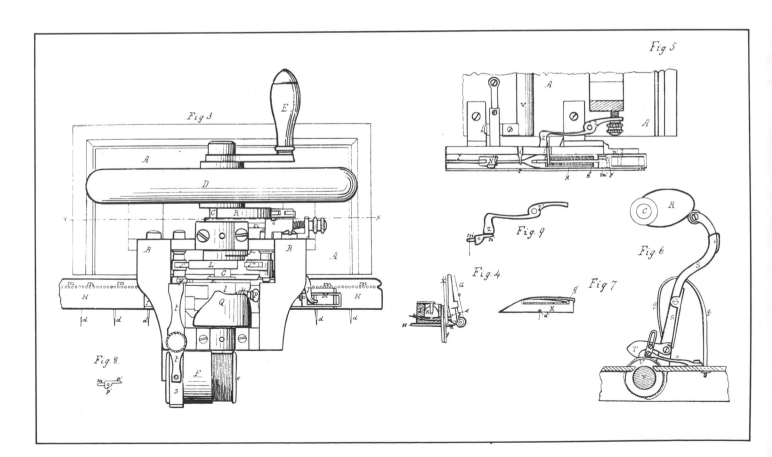

NICHOLS & BLISS,

Manufacturers of Howe's Improved Patent

SEWING MACHINES,

305 BROADWAY, NEW YORK,

33 HANOVER STREET, BOSTON.

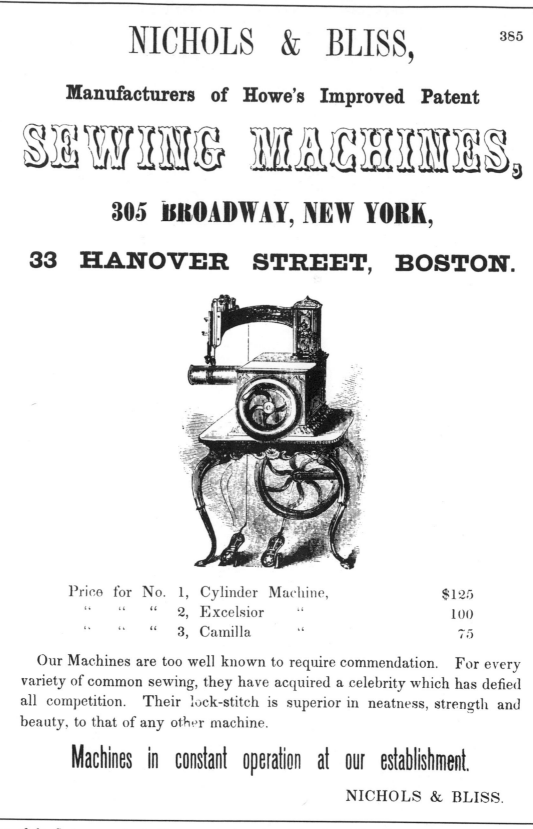

Price	for	No.	1,	Cylinder Machine,	$125
"	"	"	2,	Excelsior "	100
"	"	"	3,	Camilla "	75

Our Machines are too well known to require commendation. For every variety of common sewing, they have acquired a celebrity which has defied all competition. Their lock-stitch is superior in neatness, strength and beauty, to that of any other machine.

Machines in constant operation at our establishment.

NICHOLS & BLISS.

One of the first companies to license manufacturing rights from Elias Howe was J.B. Nichols & Co., who introduced this version of the sewing machine in 1855.

placed a machine in virtually every home in the United States and had amassed for himself a very sizable fortune.

Never one for understatement or refined taste, Singer used his new-found wealth to further the art of ostentation. He built and patented a huge horse-drawn yellow carriage, complete with nursery in the rear, that could seat thirty-one passengers. The nursery was a good idea. Singer, a bit of a philanderer, was father to some eighteen children and husband to at least two women.

On the outside of the carriage, a complete orchestra could be accommodated, with plenty of room left for the necessary guards to keep unwanted on-lookers away. Polite society, and in particular Singer's conservative lawyer partner, found it all a bit too scandalous, and Singer was quietly whisked off to France, out of the eye of his firm's more respectable clientele. Incorrigible as ever, Singer quickly attached himself to a young woman who married him and gave him six children. His final move to a colonnaded and marble-halled palace near Torquay on the English coast, where he played loving husband to his wife and doting father to all of his twenty-four children, whenever they happened by for a visit.

The first Singer sewing machine, patented in 1853.

LISHA GRAVES OTIS

VERTICAL RAILWAY

Elevators have been around for a long time. As long ago as 236 B.C., architects in the Roman Empire designed lifting devices of one sort or another, utilizing the power of men, animals, and water to operate them. With the invention of the hydraulic system of water pressure in the 1800s, European engineers were able to perfect elevators of a much more practical nature. These first hydraulic elevators were operated by means of a vertical plunger which traveled in a cylinder that extended into the ground to a depth more or less equivalent to the distance that the elevator traveled. The plunger, with the elevator car attached to its top end, moved up and down by means of the changes in water pressure governing it.

There were other types of hydraulic elevators, all employing somewhat the same principle. These elevators had their limitations, however. The maximum speed they were able to travel was 600 – 700 feet per minute, and the amount of subterranean space they required to accommodate their various mechanical riggings limited the heights to which they could ascend. It wasn't until 1899 that the first electric elevator was used commercially in the United States, installed in the Demarest Building in New York City by Otis Brothers & Company, whose founder, Elisha Graves Otis, had invented the first safety elevator back in 1852.

Otis was born in Halifax, Vermont in 1811 and worked as a master mechanic in Albany, New York. As part of his job, which included inventing labor-saving machinery, Otis was dispatched to Yonkers to construct a new factory. When

The functional aspects of the elevator was quickly hidden by opulent decoration and ornate furnishings.

he built the factory's elevator, he installed a safety device on it — an invention that would prevent the lift from crashing to the ground if the ropes that controlled it should somehow break. Within the year, he was commissioned to build two more elevators complete with safety devices. Encouraged by the prospect of a lucrative business, Otis opened a small elevator shop with his two sons and sold his first

In 1899, Otis Elevator designed quiet, vertical equipment for the 1,000-foot high Eiffel Tower. The elevator traveled at 400 feet per minute.

freight machine on September 20, 1853. Interest in the product did not come up to Otis' expectations. He received only very few orders during the first eight months.

In an attempt to drum up business, Otis demonstrated his device at an exhibition held at the Crystal Palace in New York City in May 1854. Standing on the platform of his elevator, he commanded his assistants to hoist the elevator to its maximum height. Once the platform was at its apex, Otis ordered that the hoisting rope be cut. The audience gasped. Elisha Graves Otis smiled confidently. His safety device prevented the platform from falling; he had proven that elevators were now safe enough to transport people, not just freight.

Unfortunately, the rash of orders Otis had anticipated following this daring display did not materialize, and he was forced to manufacture certain other of his many inventions, including a rotary baking oven. Once word got around about his safety device, however, business began to pick up. Hotel owners especially caught on to the possibilities of the so-called "vertical railway", department store owners and manufacturing concerns were equally enthusiastic. Otis installed the first passenger elevator in 1857 in a New York department store; two years later an elevator was installed in a New York hotel. Eleven years after that, an entire block of office buildings was equipped with an elevator. Suddenly, offices and hotel rooms with magnificent views of the New York skyline took precedence over those rooms and offices that were within easy walking distance of street level.

Although Otis died in Yonkers in 1861, the company he had founded continued, with his two sons, Charles and Norton, in control. Over the years, the Otis Brothers patented more than thirty improvements on their father's basic design. Ropes gave way to hydraulic lifts which gave way to electric lifts. In 1884 the first electric lift was installed. Five years later, at the Paris Exhibition, visitors to the Eiffel Tower were elevated 984 feet to the top of the steel structure. All of Paris was spread out before them, and the ride up had left them with more than enough breath to appreciate the view.

ELISHA GRAVES OTIS

At the Crystal Palace in New York City in May of 1854 Otis demonstrated the first vertical railway, complete with an automatic safety device. Hoisted high above the crowd, he dramatically ordered the ropes cut. The device did not fall, proving to millions of people around the world that the safe elevator had arrived.

Elisha G. Otis opened his first elevator shop on September 20, 1853, in Yonkers, New York.

In the 1870's Otis installed a steam-driven passenger elevator at New York's Lord & Taylor's department store. One important feature of this elevator was the safety hoister.

The first passenger elevator was installed by Otis in the E. V. Haughwout & Co. building in New York City in 1857. The elevator served all five stories.

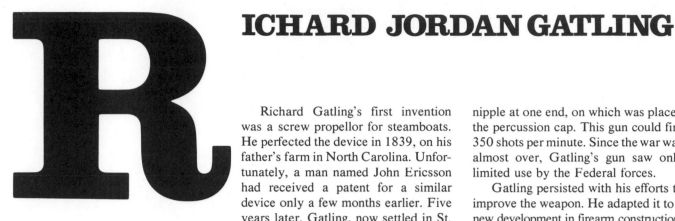

RICHARD JORDAN GATLING

MACHINE GUN

Richard Gatling's first invention was a screw propellor for steamboats. He perfected the device in 1839, on his father's farm in North Carolina. Unfortunately, a man named John Ericsson had received a patent for a similar device only a few months earlier. Five years later, Gatling, now settled in St. Louis, adapted a machine that he and his father perfected for sowing cotton so that it could sow rice, wheat, and other grains as well. His attempts were successful and the resulting machines were highly influential in transforming the country's agricultural industry.

In 1850 Gatling invented a hemp-breaking machine, and seven years later came out with a practical steam plow. When war broke out between the North and the South, he immediately turned his attention to the development of firearms. Several examples of rapid-fire machine guns were being used during the early years of the war. At the battle of Fair Oaks, in Virginia, in May of 1862, Confederate soldiers used a weapon called the Williams machine gun, which marked the first time a machine gun had been used in warfare. Other common guns at the time were the Ager "coffee mill" gun, otherwise known as the Union gun; the Ripley machine gun; the Claxton; and the Gorgas machine guns.

Gatling examined all these makes, and, borrowing heavily from the advanced features of the Ripley and Ager models, fashioned his original gun in 1862. His version had a single-barrel device with a rotary chamber, and used the best ammunition available. This consisted of a paper cartridge placed within a heavy-walled steel cylinder that was countersunk at the base. The cartridge itself was fashioned with a nipple at one end, on which was placed the percussion cap. This gun could fire 350 shots per minute. Since the war was almost over, Gatling's gun saw only limited use by the Federal forces.

Gatling persisted with his efforts to improve the weapon. He adapted it to a new development in firearm construction: a lightweight brass cartridge that contained, in a single unit, a means of ignition, the propellant, and a bullet. This advance completely changed the design of all future firearms. Although it was not Gatling's invention, he was one of the first to use the cartridge in a machine gun. His new design consisted of several barrels mounted parallel to each other and spaced at equal distances to rotate around a central shaft. A hand crank that could be turned either slowly or rapidly activated the gun, thereby providing optimum control of the weapon by the user. A combination of gravity and the camming action of the cartridge container positioned the incoming cartridge which was loaded at the same time that the barrel rotated around the shaft. The device fired at mid-revolution and extracted and ejected the cartridge immediately thereafter.

Once perfected, Gatling's gun was the best manually operated one around. When powered externally, it was capable of firing 3,000 rounds per minute. One disadvantage to the weapon, particularly for tactical use, was its reliance on a multibarreled arrangement. When inventors tried to improve on the Gatling machine gun, that was the area on which they first concentrated.

Gatling's invention had gained him a sizable fortune and worldwide recognition. He died in New York City in 1903 a rich and very famous man.

RICHARD JORDAN GATLING

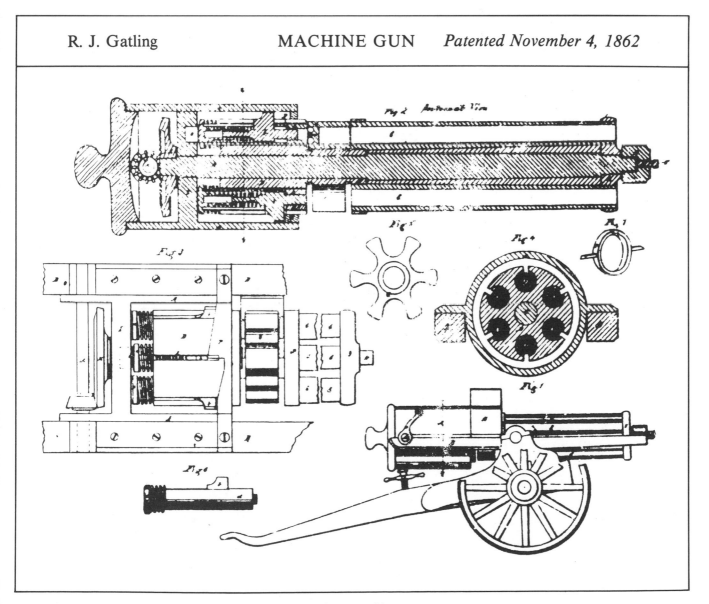

R. J. Gatling　　　　MACHINE GUN　　　*Patented November 4, 1862*

Illustration from Richard J. Gatling's patent application for a machine gun.

HENRY BESSEMER
WILLIAM A. KELLY

STEEL

Holding ownership of a patent does not necessarily ensure the end to your troubles. This is a fact that most inventors discover somewhere along the way. For every person with a patent there is another ready to dispute its legality. In the case of the inventors of steel, what had begun as just another bitter rivalry turned into a very happy and profitable financial merger for all of the parties involved.

English inventor and engineer Sir Henry Bessemer is credited with the discovery and development of the first process for making abundant and inex-

pensive steel by taking the carbon out of molten pig iron. Born in 1813 in Charlton, England, Bessemer was essentially a self-taught engineer. He devised improvements on typesetting machines and made a small fortune by producing gold-colored powder, used for making gold-colored paint.

During the Crimean War, while working on the development of a rotating artillery shell, Bessemer saw the need for a stronger metal with which to make a cannon that would take his shell. Through experimentation, he developed the process that has since been known

Hand Lathe.

Henry Bessemer invented the converter while working on a rotating artillery shell during the Crimean War. He used as his guide several other rotating machines, including some inventions by Leonardo da Vinci.

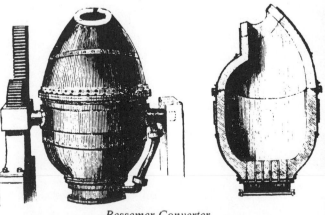

Bessemer Converter.

58

as the Bessemer converter. By blowing air through molten pig iron, Bessemer was able to remove the carbon from the substance, while at the same time maintain sufficient heat to keep the material hot and liquid. The resultant metal was an extremely malleable mild steel.

With the help of Goran Goransson, a Swedish licensee, Bessemer developed improvements and was able to market the material with great success. He was awarded two patents by the United States Patent Office in 1865, the first on February 12, the second on August 25. It was at this point that William Kelly entered the scene.

Kelly was born in Pittsburgh in 1811. In partnership with his brother, he purchased iron-ore land and a furnace in Kentucky, where he manufactured sugar kettles. Concerned with the deple-tion of his fuel supply, which at that time was exclusively charcoal, Kelly happened on a fortuitous discovery. He noticed that the molten iron in his furnaces, when not covered by the charcoal fuel, would burn to a white heat whenever an air blast hit it. He quickly discovered that the carbon in the iron acted as a fueling agent and, in the process of being burned out, caused the molten material to become even hotter.

Between 1851 and 1856, Kelly constructed seven experimental converters, which he worked on in secret. In June 1857, by somehow convincing patent officials that his was the priority claim, William Kelly received a patent for his invention, which was the same as Sir Henry Bessemer's. In 1863 the Kelly Pneumatic Process Company was formed

HENRY BESSEMER

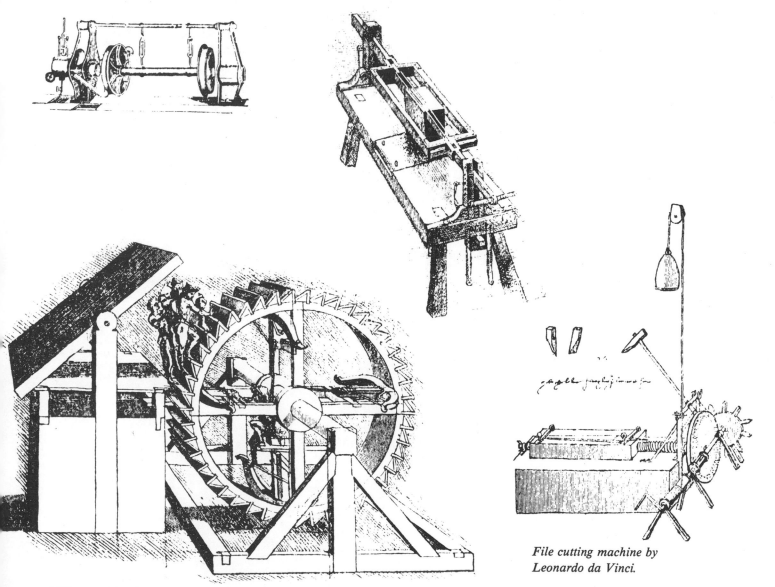

Military engine by Leonardo da Vinci.

File cutting machine by Leonardo da Vinci.

to market the converters in the U.S.

Meanwhile, a man named Alexander L. Holley, learning about the Bessemer process while on a trip to England and impressed by it, acquired United States rights to the Bessemer process and returned to America determined to market it there. Enlisting the help of his two associates, John A. Griswold and John F. Winslow, Holley constructed a processing plant in Troy, New York where he improved upon the Bessemer process.

By the beginning of 1865, there were two groups of steel processing patent holders, each convinced of its legal right to the process, and each financially capable of marketing it. Thus the battle began. Fortunately — and it was virtually without precedent among similar patent fights — the two parties came to terms. After investing huge sums of money in legal fees, and one year in court battles, in 1866 the two groups combined their respective interests. Winslow, Griswold, and Holley received seventy percent and the Kelly Pneumatic Process Company thirty percent of all subsequent royalties on the invention.

An important part of making iron was filling the blast furnace.

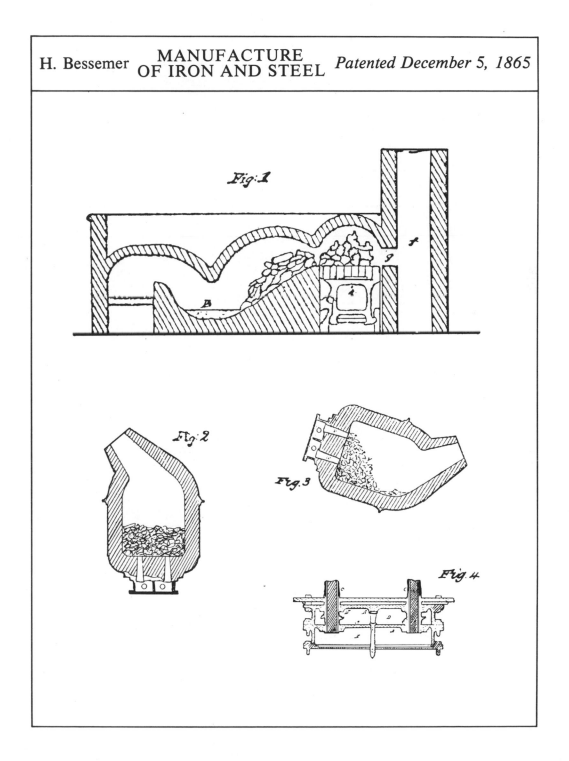

Fig. 1

Fig. 2

Fig. 3

Fig. 4

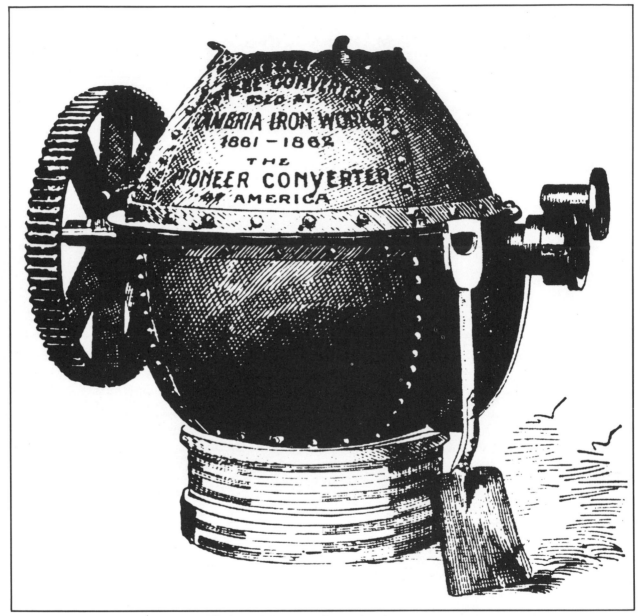

Converter used at the Cambria Iron Works in Johnstown, Pennsylvania.

Henry Bessemer's version, in action.

EORGE WESTINGHOUSE, JR.

STEAM
POWER BRAKE

George Westinghouse received his first air-brake patent in April 1869. Six months later he formed the Westinghouse Airbrake Co. to manufacture the equipment. Over the years, as he continued to make improvements and adjustments on his initial invention, he was awarded more than twenty additional patents for air-brake equipment alone.

Westinghouse received his mechanical training while working in his father's agricultural implement factory following service in both the navy and the army during the Civil War. The very first patent he received was for a rotary steam engine in 1865; in the same year he patented a device that put derailed freight cars back on the track. But it was the invention of the air-brake that was of major importance in developing rail traffic. Westinghouse's invention assured the safety of high-speed travel by rail. It was 1872 before he manufactured an automatic air-brake, which he sold almost immediately to both European and American railway companies.

Over the course of his lifetime, Westinghouse secured more than 400 patents for such diverse inventions as electrical power transmission and railway signals. As president of the Westinghouse Electric Company, he built dynamos for the rapid transit system in New York City and the London Metropolitan Railway. In addition to acquiring patents for his own inventions, Westinghouse bought the patents for signal and interlocking switch devices invented by others, which he combined with his own inventions to make a complete signal system. While head of his own Union Signal and Switch Co., Westinghouse developed an interest in more generalized electrical processes. He introduced alternating current for electrical power transmission in the United States and formed the Westinghouse Electric Corporation, which became one of the largest electrical manufacturing companies in the country.

Intrigued by the advancements made by Europeans who experimented with electricity, Westinghouse purchased test equipment and hired a team of engineers to adapt the European system of lower electrical voltage by way of transformers for use in America. William Stanley was one of the engineers hired. In 1886, several buildings in Great Barrington, Massachusetts were illuminated by means of Stanley's system of high tension transmission. Similar success was achieved in a suburb of Pittsburgh. On the basis of this publicity, Westinghouse received the contract to light the Chicago World's Columbian Exposition in 1893, as well as to develop the power of Niagara Falls.

Until Westinghouse lost control of his companies in 1907, his business interests flourished. He was reinstated as president when the companies were reorganized, but his powers were greatly reduced. He stayed on, however, until 1911, also serving as a trustee of the Equitable Life Assurance Society from 1905 to 1910. Throughout his professional career he continued with his experimentation. Even after he had retired, he worked on the steam turbine, and reduction gear, used in speed & power adjustments, as well as an airspring for automobiles. Gradually, his health worsened, and he died in 1913.

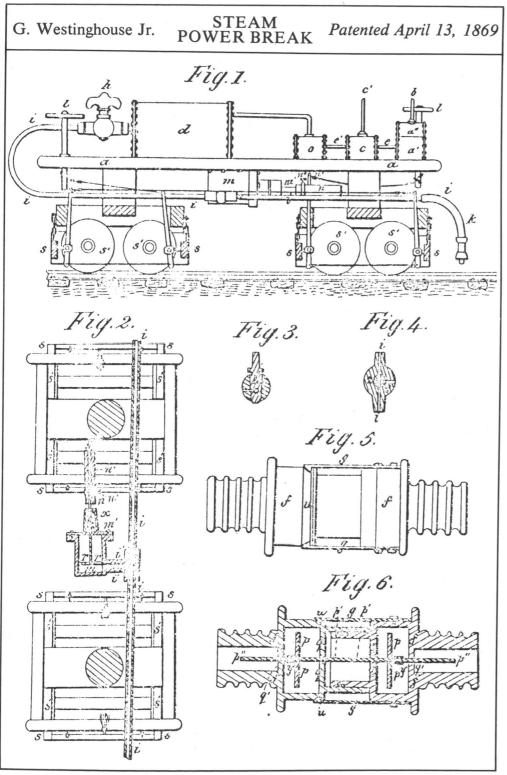

Fig. 1.

Fig. 2.   *Fig. 3.*   *Fig. 4.*

Fig. 5.

Fig. 6.

The patent application for George Westinghouse "Steam Power Brake"

GEORGE WESTINGHOUSE

OSEPH F. GLIDDEN

BARBED WIRE

The inspiration for barbed wire came from an itinerant salesman in the midwest, who demonstrated a new type of fencing at the DeKalb county, Illinois fair in 1873. Along the wire of an ordinary fence, the salesman attached a long strip of wood, from which protruded several sharp spikes. To the astonishment of the spectators, he showed how cattle, afraid of the sharp spikes, would not venture near the fence. The Illinois farmers were fascinated. Situated on the eastern brink of the American plains, these men were constantly on the lookout for better ways to work their farms. Good fences were difficult to come by in the West:

wood was scarce, there weren't enough rocks to build stone walls, and strands of wire could not hold up against the extreme temperatures or the roving cattle.

Joseph F. Glidden was one DeKalb farmer who saw possibilities in the fences the salesman was demonstrating. He immediately went home and set to work perfecting a spiked fence of his own. By twisting two strands of wire together and tying a short piece of wire between the twists, Glidden happened upon barbed wire. He quickly patented his invention, followed almost immediately by another Illinois resident, Jacob Haish, and then Isaac Hellwood,

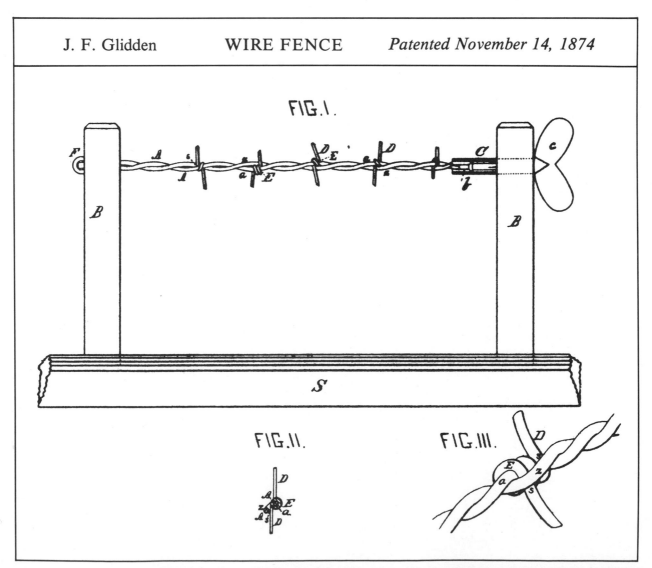

Illustration from J.F. Glidden's patent application for barbed wire fences.

who had invented separate barbed wire versions of their own. Thus began eighteen years of bitter court battles which, in the end, were won by Glidden.

Glidden started his barbed wire construction at home, twisting the wire on his grindstone and fashioning the barbs with his wife's coffee mill, but he soon moved into a factory in DeKalb. Barbs were slid onto the wires by young boys, then transported to the factory where they were spaced and twisted. Glidden ordered his wire from a manufacturing concern in the East. Intrigued by the large orders coming from DeKalb, the company dispatched an employee to investigate the situation. He brought back samples of Glidden's wire. Immediately, the company hired an inventor to devise a machine for making barbed wire automatically. Once that was accomplished and the machine patented, the company bought Glidden's business, increasing the output to 600 miles of wire per ten-hour day. In 1876, the automatic wire maker produced the fencing at the rate of seventy barbs per minute. It was now only a question of marketing.

Salesmen traveled west with the wire, demonstrating how quickly the wire could be strung, and how easily it could be attached to its supporting posts. Pretty soon, shipments of barbed wire were arriving by the trainloads, to be used by ranchers and farmers alike. Two of Glidden's own relatives were among the vanguard of these barbed wire salesmen. In 1880, Glidden's son-in-law, a Mr. Bush from Chicago, and H.B. Sanborn established a cattle ranch in the Texas panhandle. Rather than allow the cattle to roam free, they fenced in all 150 miles of their Frying Pan Ranch, to the tune of $39,000. This signaled the end of the free range, and the beginning of a brief but heated battle

between fencemen and no-fencemen, between farmers and cattlemen. Range wars broke out all over Texas, Arizona, Wyoming, and Oklahoma. Obviously, the fencemen won.

It wasn't long before man devised other uses for the wire. In the Boer War in South Africa, barbed wire received its introduction to real warfare. During World War I, the front lines were etched with it. And the Berlin Wall, surrounded by hundreds of yards of barbed wire, has become a symbol of persecution and oppression to half mankind. The simple invention designed to benefit farmers and cattle ranchers has certainly come a long way from that county fair in De-Kalb, Illinois, back in 1873.

A great dispute arose between farmers and cattle ranchers over the use of barbed wire. The farmers wanted to fence in their property to protect their crops and contain their animals; the cowboys wanted the range to be free. Range wars broke out all over the Southwest.

"I wish the man who invented barbed wire had it all wound round him in a ball and the ball rolled into hell." Fence-cutter wars are not often mentioned in cowboy films. But they happened in the real American West less than one hundred years ago. Men drew their guns over barbed wire. Every roll of the ferocious-looking stuff, painted a sticky black or sometimes bright red, meant land fenced in that had never been fenced in before.

The Indians called it the "Devil's Rope"

HELL BROKE LOOSE IN TEXAS

Wire-cutters Cut 500 Miles in Coleman County

The enemies of barbed wire crept out quietly in the night. A few snips here, a few snips there, and miles of newly strung fencing came down. The rancher who had just paid to put it up fought back. "The fence-cutters themselves have told me," said a Texas Ranger in 1888, "that while a man was putting up his fence one day in a hollow, a crowd of wire-cutters was cutting it behind him in another hollow back over the hill." It was an angry Texan rancher who wished the inventor of barbed wire in hell.

There was war between the inventors of barbed wire, too — they fought in the law courts. Each one said he got the idea first. The Great Barbed Wire Case lasted eighteen years, before the United States Supreme Court decided that old Joseph Glidden of De Kalb, Illinois, was the inventor.

But Jacob Haish, also of De Kalb, built himself a big expensive house and carved over the door, JACOB HAISH. *Inventor of Barb wire.*

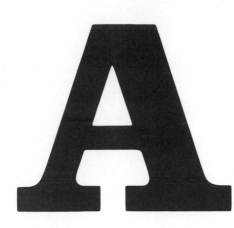

TELEPHONE

ALEXANDER GRAHAM BELL

The man who is credited with the invention of the telephone developed an early interest in voice and speech patterns and hearing and vocal vibrations from his father, Alexander Melville Bell, and his grandfather, Alexander Bell, both noted authorities on phonetics and speech therapy.

Born in Edinburgh, Scotland in 1857, Alexander Graham Bell studied at the University of Edinburgh, where his father was a lecturer on elocution, and then moved with his family to London in 1867, where he entered London University. Ill health forced him to leave England and move to Canada, but before he left he became deeply involved in work with the deaf, adapting his father's system of "Visible Speech" — which provided symbols for the position of the vocal chords in speech — for use in teaching the deaf to talk.

Once in North America, Bell soon settled in Boston, where he opened a school for the training of the deaf, sponsored by the Boston School Board. Within a year he was appointed professor of vocal physiology and the mechanics of speech at Boston University. During this time, he continued to instruct students privately, as well as work on several of his projects, among which were a multiple telegraph, a phonautograph, which produced writing or a tracing by means of sound vibrations, and an electric speaking telegraph or telephone.

Early in his career, Bell had been associated with Charles Wheatstone, the man who invented the telegraph, and Alexander John Ellis, an expert on sound. Ellis had shown Bell how an electric current could affect the vibration of a tuning fork. Bell incorporated this discovery into his experiments with

Although Bell's telephone was protected by the patent law, other inventors worked to improve on his original design. Elisha Gray's 1882 version.

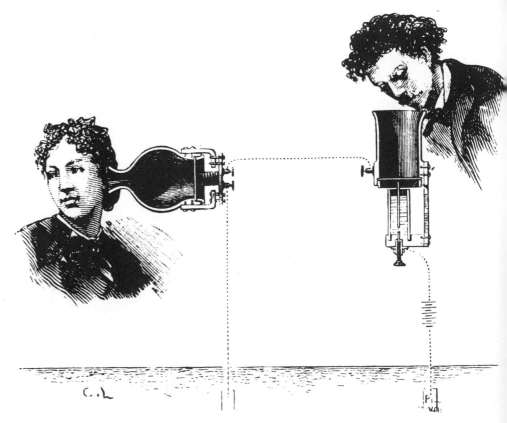

the human ear. In 1874 he had conceived the theory of the telephone; in 1876, through the encouragement of Professor Joseph Henry, he finally resolved the problem of transmitting the voice electrically.

By constructing artificial eardrums out of thin metal strips and then wiring them with electric wire, Bell devised an apparatus that would vibrate when contacted by sound and in turn would cause a change in the magnetic field surrounding it. The electric currents produced within the apparatus corresponded to the sounds that were transmitted into it. It wasn't long before Bell was able to pass a vocal message through his invention to an assistant in an adjoining room. The world's first telephone conversation was transmitted between Boston and New York only one year after Bell exhibited his invention at the 1876 Centennial Exhibition in Philadelphia.

Bell secured two separate patents for his invention, in 1876 and 1877, and quickly formed the Bell Telephone Company in order to develop them commercially. In 1878 the first switchboard was installed, in New Haven, Connecticut. Meanwhile, Bell had taken his new bride, Mabel G. Hubbard, a woman who had been deaf since early childhood, on a honeymoon trip to Europe, where, in 1877, he introduced his invention to England and France. Upon his return to the United States, he continued to perfect his telephone.

In 1882 he became a United States citizen, and worked on his many other inventions, among them an audiometer, an instrument to engage and record the acuteness of hearing, and several improvements on Edison's phonograph. In 1880, in cooperation with Sumner Taintor, he had completed his preliminary work on the photophone, a machine that produced and reproduced sound by means of light waves, and now he developed his first working model.

Toward the end of the century, while Bell was president of the National Geographic Society, he turned his attention away from audio equipment to airplanes. He encouraged experimentation in the field of aviation, and invented the tetrahedral kite, which was used in the study of wind and air currents. In 1907 he

ALEXANDER GRAHAM BELL

Bell traveled all over the world to publicize his telephone. In 1877, he delivered a speech to an audience, fifteen miles away.

founded the Aerial Experiment Association, which financed the first public flight of a heavier-than-man air machine in 1908. Bell maintained his enthusiasm for the possibilities of air travel. He continued to work in his laboratory, sit at the piano playing old Scottish tunes, and entertain his family with recitations of Shakespeare until he died in 1922 at the age of 65, in his summer home in Baddek, Nova Scotia.

The Bell Telephone Company was founded in 1877. Soon telephones were being used throughout the country for business and pleasure.

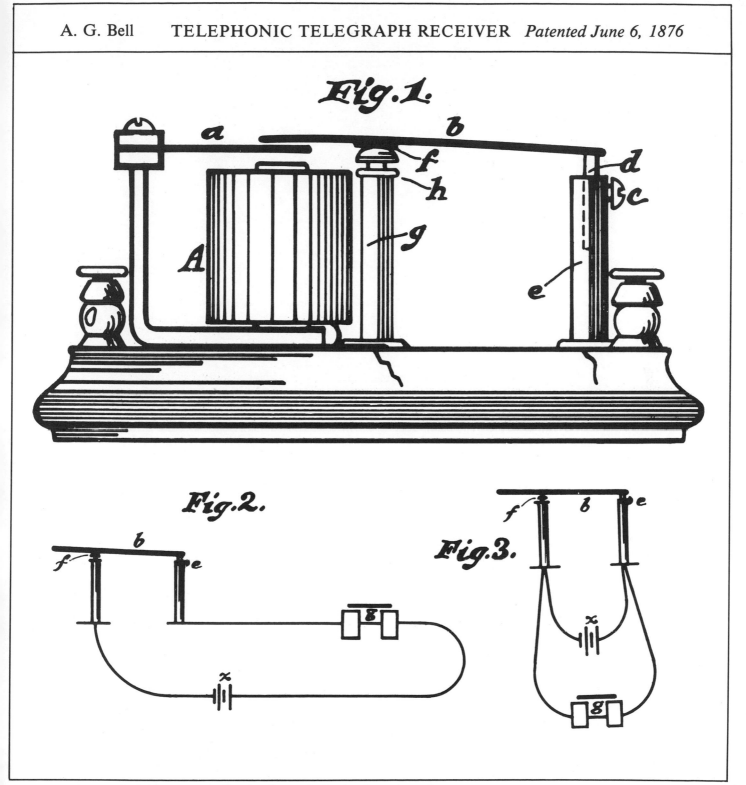

Fig. 1.

Fig. 2.

Fig. 3.

Illustration from A. G. Bell's patent application for a telephonic telegraph receiver.

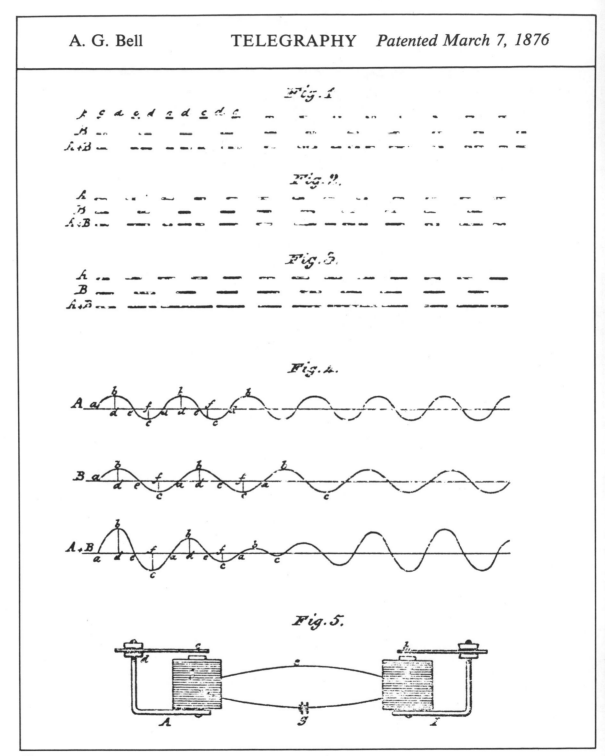

Illustrations from A. G. Bell's patent application for telegraphy.

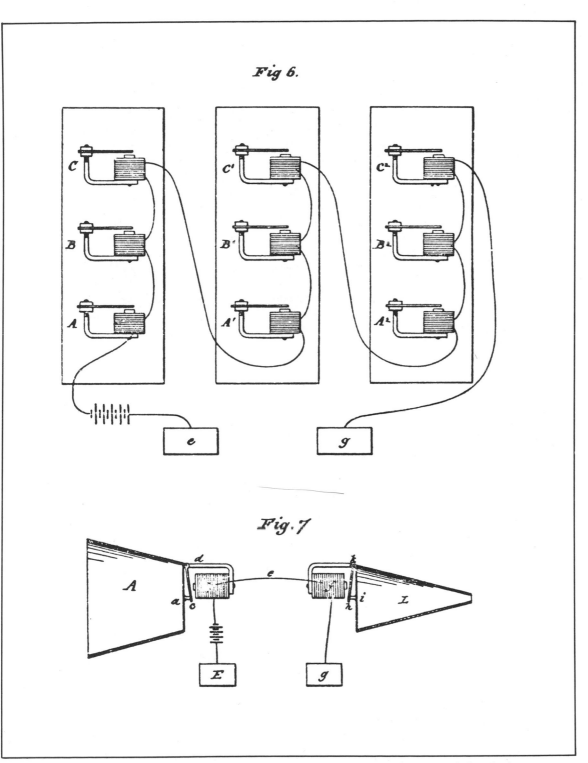

Fig 6.

Fig. 7

Bell's telelgraph, illustrated above, was patented only a few months before he patented the telephone.

NIKOLA TESLA

ELECTRO MAGNETIC MOTOR

Nikola Tesla, a Croatian by birth, emigrated to the United States in 1884 and worked for a time with Thomas Alva Edison's company. Born in 1856, Tesla had studied physics and mathematics at the polytechnic school of Graz and the University of Prague. Like Edison, Tesla had an intense interest in electrical engineering and upon leaving Edison's employ, he established the Tesla Institute in New York City where he devoted all his time and energies to electrical research.

While working with Edison, Tesla developed and patented the induction, synchronous, and split-phase motors, as well as new forms of generators and transformers. George Westinghouse used Tesla's system of alternating current to light the World Columbian Exposition at Chicago in 1893. As a result of this success, Westinghouse received a commission to convert Niagara Falls for electric power usage; The power machinery that he used bear Tesla's name and patent numbers. By 1896, the city of Buffalo was receiving power from Niagara Falls.

Tesla's first discovery was that the rotating magnetic field principle could be used as an effective method of alternating electric current for power. In a series of lecture tours throughout Europe and the United States, he demonstrated several discoveries and applications of high-frequency alternating current, including the high-frequency resonant transformer which bears his name, the Tesla coil.

Within his lifetime, Tesla was recognized as one of the foremost specialists in electrical technology. His invention of the electromagnetic motor, patented in 1888, was instrumental in converting electrical energy into mechanical energy. It was Tesla's idea to explore the interaction of stationary and moving parts within the motor itself. His electromagnetic device, more than any other electrical-mechanical system, was able to maximize the amount of energy converted in relation to unit weight, volume, and cost.

Tesla constructed his motor in essentially two parts: a rotor and a stator. The rotor was the movable part; it contained conductors which established and shaped the magnetic fields that came into contact with the magnetic fields surrounding the stator. A contacting device within the rotor connected it electrically with the stator. In addition to the magnetic and electrical conductors, which were similar to those of the rotor, the stator also served as the frame which supported the entire device.

Tesla's development of the electromagnetic motor and subsequent discoveries and accomplishments in all areas of electrical research earned him international acclaim. In 1917 he was awarded the Edison Medal, the highest honor that the American Institute of Electrical Engineers could bestow. Although he died in New York City on January 10, 1943, his reputation in scientific circles has increased. In 1956, at a celebration of the 100th anniversary of Tesla's birth, a unit of magnetic flux density in the MKS system was named "tesla" as testimony to the accomplishments of this great man.

NIKOLA TESLA

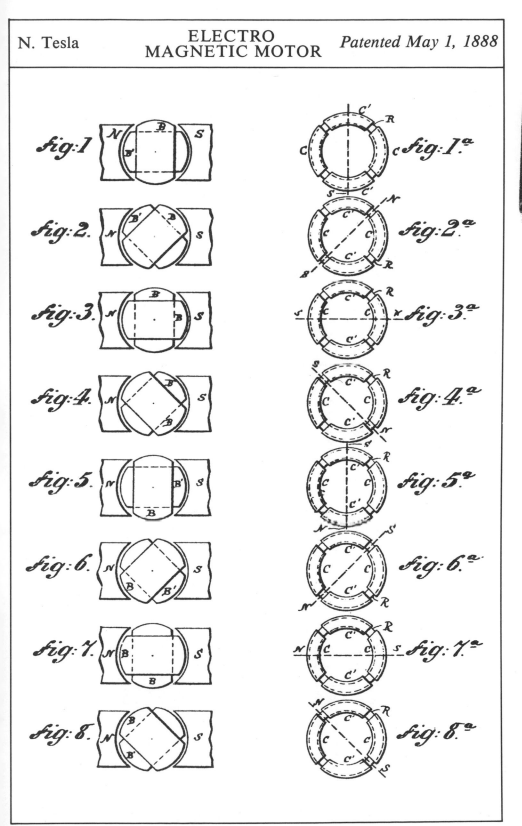

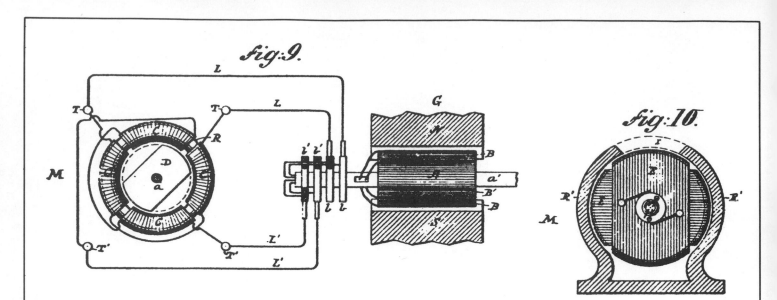

Fig:9.

Fig:10.

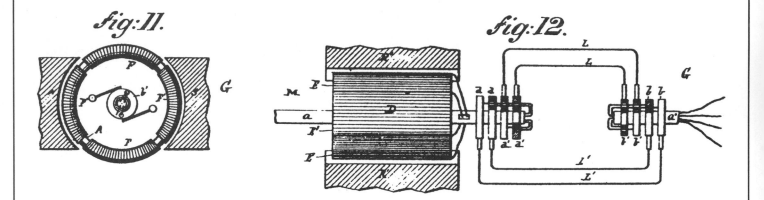

Fig:11.

Fig:12.

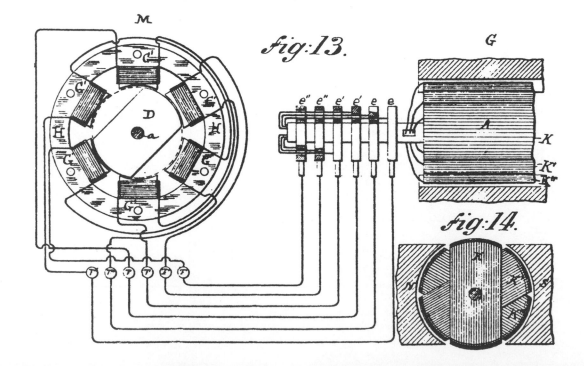

Fig:13.

Fig:14.

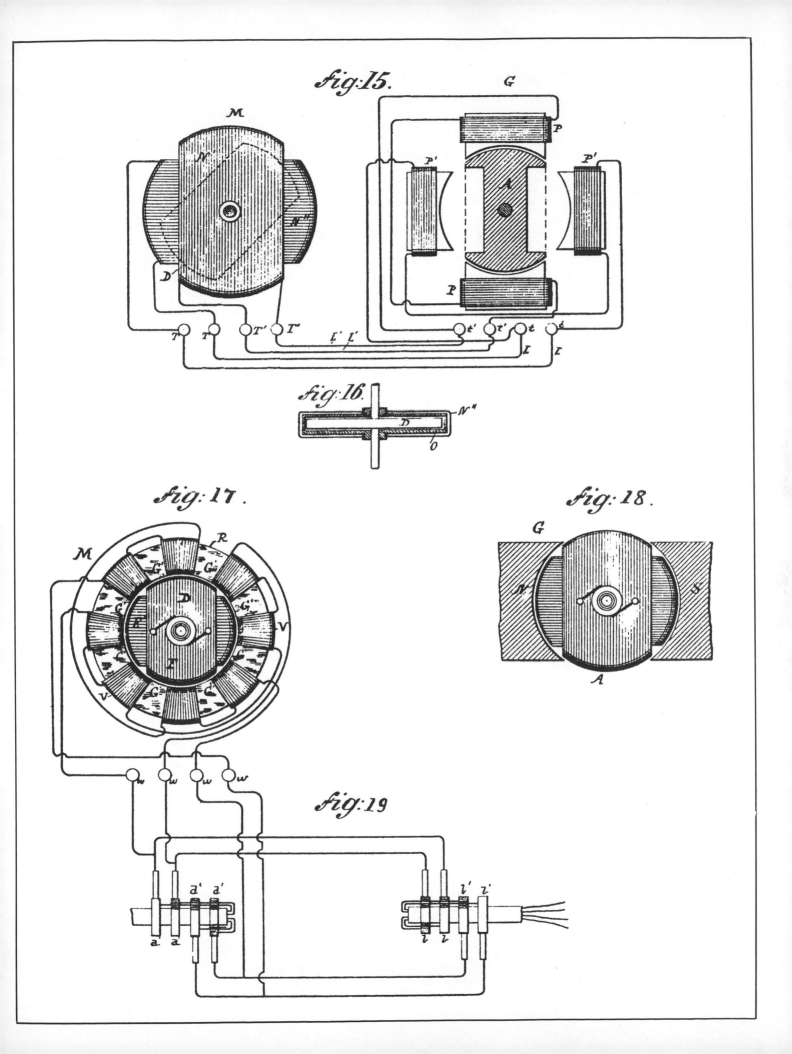

Fig:15.

Fig:16.

Fig:17.

Fig:18.

Fig:19.

JOHN J. LOUD

PEN

When John Hancock signed the Declaration of Independence, he used a quill feather to write his name. It was the only practical instrument for writing by hand that had been invented since the Middle Ages, when monks started using feathers from swans or turkeys to write their manuscripts. The nineteenth century changed that. In 1803 a steel nib pen was invented. Although the actual inventor is unknown, John Mitchell, Joseph Gillott and Josiah Mason were early leaders in the manufacture of steel pens. By 1830, the nibs were being manufactured by machines in Birmingham, England. Fifty years later, in New York City, Lewis Edson Waterman invented the first fountain pen. The pen's stem contained a small reservoir of ink which was placed inside with an eyedropper.

In 1888 John J. Loud became the

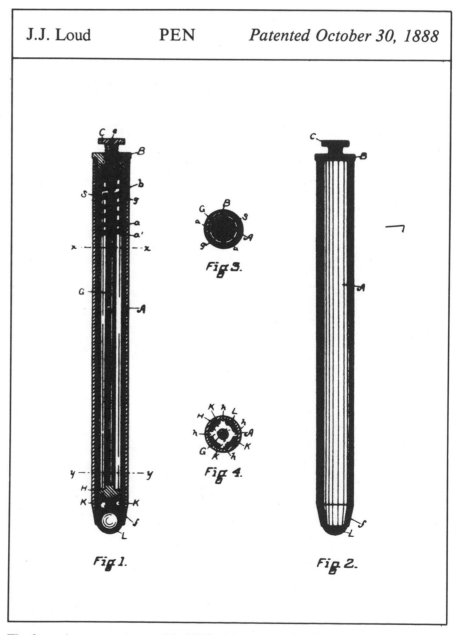

The fountain pen was invented in 1880, eight years before John J. Loud patented his ballpoint pen.

first person to patent a ballpoint pen. Unfortunately, Loud's invention did not take the country by storm; it was something of a commercial failure. People were more interested in the development of the fountain pen, which had proven to be such a welcome advance over the use of feathers. Early in the twentieth century, the self-filling fountain pen was introduced onto the market. Equipped with a lever on the shaft, the pen could be filled by dipping the nib into the ink and then raising and lowering the lever. The rubber reservoir in the pen itself was thus deflated and inflated and the ink drawn into it by suction. The quill feather was definitely obsolete.

With the success of the fountain pen, inventors turned their attention back to Loud's achievements. In the 1930s, two Hungarian brothers, Georg and Ladislao Biro, invented what is considered the modern ballpoint pen. But, like Loud, the Biro brothers did not meet with instantaneous success in marketing their invention. It took ten years for the pen to become freely available to the public. A factory was established in England in the mid-forties which mass-produced them. Since then, of course, the ballpoint pen has become such a universal item and so inexpensively manufactured that it has outdistanced the fountain pen which for so long maintained an advantage over it. But the past few years have shown that the ballpoint pen itself may soon be eclipsed by the felt tip pen, which has enjoyed a booming reception on the American market.

Little did John J. Loud realize, way back in 1888 when he tried unsuccessfully to market his invention, that the principles inherent in his first design would have such extensive commercial possibilities, or that it would take well over half a century for the public to finally catch on.

"Will you please stop retracting your ball-point pen?"

W

ZIPPER

HITCOMB L. JUDSON

Whitcomb L. Judson was a restless inventor. Within a sixteen-year span he was awarded thirty separate patents.

On the day Lewis Walker, a Pennsylvania businessman, arrived in Minneapolis, Judson was discussing a hot new idea with James R. Williamson, a patent attorney. Judson had designed a railway system run on compressed air. He had already completed two successful experimental operations of his idea, in Washington, D.C. and in New York City, and corporations from all over the country were in the midst of bidding on his transportation system. Walker, who was present at the meeting, was excited by Judson's idea, but before he could act on it, the plan fell through. The sudden availability of plentiful amounts of electric energy made the wide-spread use of compressed air propellants obsolete.

Judson turned his energies to new ideas and inventions. On August 29, 1893, he was awarded two patents on

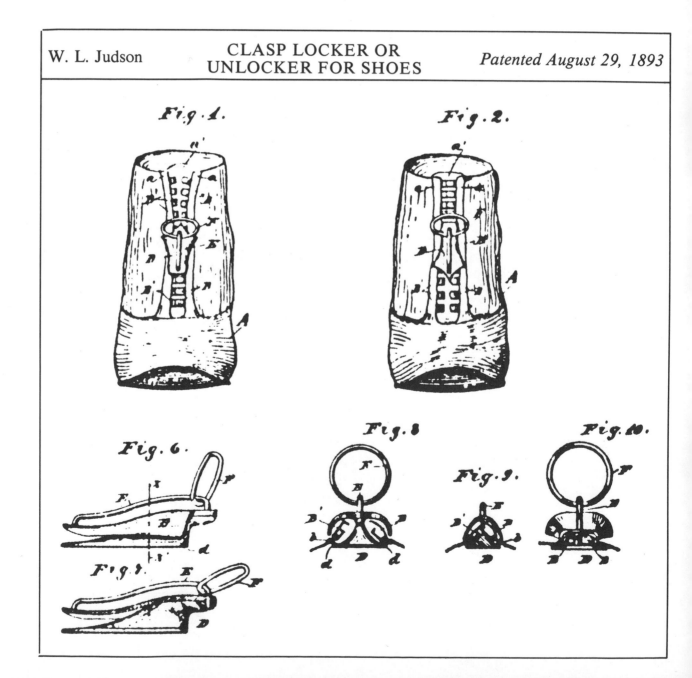

| W. L. Judson | CLASP LOCKER OR UNLOCKER FOR SHOES | *Patented August 29, 1893* |

his latest invention – what he called a "clasp-locker," which he used to fasten together his boots. Judson showed the locker to his partner, Harry Earle, in New York, and then to Lewis Walker, who was very enthusiastic. Walker immediately went out and had a pair of his own boots fashioned with the new device and wore them everywhere, convinced that this was a great invention.

The three men formed a partnership in 1894, called the Universal Fastener Company and based in Chicago. Judson continued to make improvements on the fastener and took out two more patents on it on March 31, 1896.

Although the three men were certain that the fastener was a good idea, they were puzzled as to how to go about marketing it. Because the fastener was

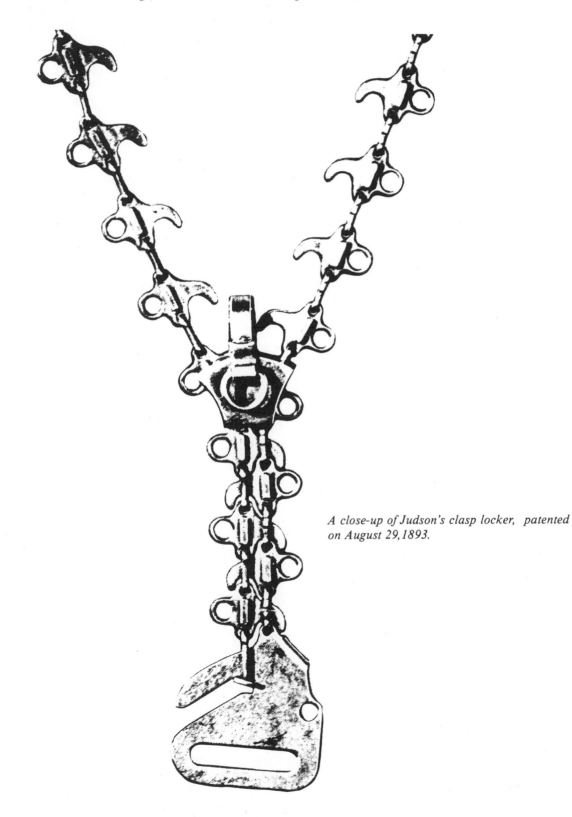

A close-up of Judson's clasp locker, patented on August 29, 1893.

heavily constructed, commercial uses seemed at first to be the most appropriate. The company received an order from the United States Postal Service for twenty mail bags equipped with the new device, and for a while the company's prospects seemed good. Lucrative contracts, however, did not come its way.

Alternative uses, such as for leggings, were considered, but business failed to improve and the company's investors withdrew their money. Walker, however, continued his support. Moving the company to Elyria, Illinois, he continued to investigate new marketing areas. Judson was put to work designing equipment to manufacture the device, and Earle went to Europe in an attempt to market the fasteners abroad. The political and economic climate in England in November of 1898 was not conducive to interesting financial investors in foreign enterprise. Several business scandals had recently surfaced, culminating in some rather sensational court trials. Earle returned home dejected; in the interim, another of Judson's inventions – a machine to manufacture the fasteners automatically – collapsed in the midst of its maiden run. The fastener business came to a halt.

Four years later, in 1905, Judson secured yet another patent. This one was for a fastening device called "C-Curity." In his experimentation, Judson came up with the idea of attaching the fastener to a piece of cloth, called a placket; all previous designs had involved individual attachment of the metal hooks and eyes to the garment itself. Judson's invention brought the development of the zipper one step closer to universal acceptance.

Inventive as he was, however, Judson was not a mechanical genius, and the C-Curity fastener was not very reliable. It wasn't until 1906, when Gideon Sundback was engaged to turn Judson's design into a practical commodity, that the fastener finally got off the ground.

Sundback struggled through several years of frustrations and disappointments before he applied for a patent on an entirely new kind of fastener on October 22, 1912. He received it in 1917. In 1914, the first plant for the successful manufacture of zippers in the United States was opened in Meadville, Ohio. Among the men involved in the plant's creation was Lewis Walker, who had maintained his support and enthusiasm for the idea since he first heard about it back in 1889.

Whitcomb L. Judson, the man with the idea, was not a part of the happy occasion in Meadville. He had died in 1909, unaware of the effect his "clasp locker" was to have on the generations ahead.

In 1905, Judson's idea turned into a practical reality. Named "C-Curity," this workable zipper was marketed to a female audience.

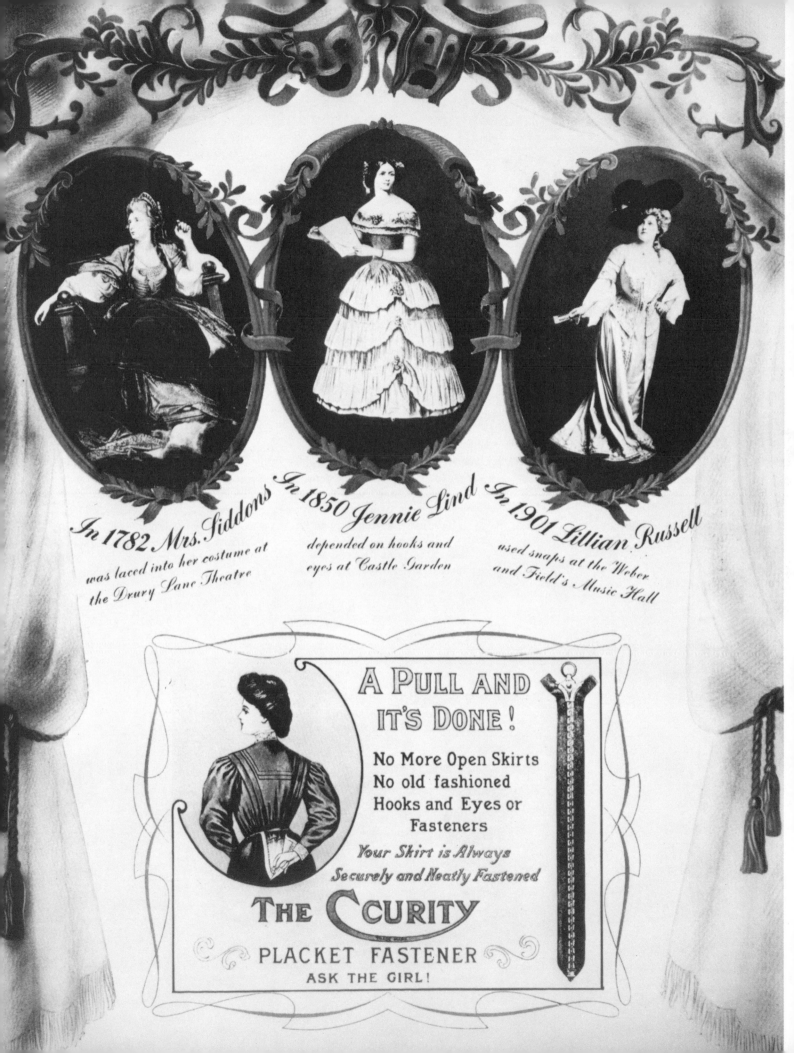

In 1782 Mrs. Siddons was laced into her costume at the Drury Lane Theatre

In 1850 Jennie Lind depended on hooks and eyes at Castle Garden

In 1901 Lillian Russell used snaps at the Weber and Field's Music Hall

A PULL AND IT'S DONE!

No More Open Skirts
No old fashioned
Hooks and Eyes or
Fasteners

Your Skirt is Always Securely and Neatly Fastened

THE CCURITY

PLACKET FASTENER
ASK THE GIRL!

GUGLIELMO MARCONI

TELEGRAPH

Like most inventors, Guglielmo Marconi initially had a difficult time convincing the public that his discovery was important. Yet his system of radio-telegraphy had incredible possibilities for practical application. Unappreciated in his native land, Marconi had to move to England to receive the encouragement he needed to publicize and market his work.

Born in Bologna, Italy in 1879, the son of an Italian country gentleman and an Irish mother, Marconi studied physics at the technical school in Leghorn. It was there that he became familiar with the work on electromagnetic waves that had been developed by James Clerk Maxwell, Heinrich Hertz, Edouard Branly, Sir Oliver Lodge, and Augusto Righi.

When he was twenty, Marconi started his own experiments. Using his father's estate in Pontecchio, close to Bologna, as a workshop, he constructed a relatively crude machine for transmitting signals without wires. This machine consisted of an induction coil with a spark discharger that was controlled by

a Morse key at the sending end and, at the receiver, a simple coherer. The coherer, which is essentially a small glass tube loosely filled with iron filings that cohere and form a conductor when electromagnetic waves are passed through them, was invented by the French electrician Edouard Branly and proved to be a tremendous aid to Marconi in his experimentation.

Marconi made improvements on Branly's invention, substituting nickel and silver for the iron filings and using silver plugs to close the tube. With this crude machine, he was able to increase the range to a distance of a mile and a half. This success convinced him that he had effected a viable system of communication. Unfortunately, there were not too many other people in Italy at the time who shared his enthusiasm or saw the machine's potential.

Therefore, in February 1896, Marconi moved to London. William Preece, who at that time was engineer in chief of the British post office, was very excited about Marconi's invention, and, with Preece's assistance and encouragement,

Telegraph office at Sandy Hook.

Marconi filed his first patent in June 1896. He quickly began a series of very successful demonstrations using kites and balloons to obtain greater heights for his aerials, thus being able to transmit signals over a much greater distance – up to four miles across Salisbury Plain and nine miles across the English Channel. These demonstrations, as well as Preece's lectures about them, attracted a great deal of publicity throughout the world.

Within a short time, Marconi was able to transmit signals from a communication center at La Spezia to Italian warships as far as twelve miles away. This was followed by a land station at South Foreland, England, which was able to communicate over a distance of thirty-one miles with a similar station in Wimeroux, France. In the same year, messages were transmitted between British battleships as far as seventy-five miles apart.

With the help of a cousin, Jameson Davis, Marconi formed Marconi's Wireless Telegraph Co., Ltd., in 1900. The American Marconi Company was established shortly thereafter, due primarily to the success achieved when radio equipment was installed in two United States ships in order to report the progress of the America Cup yacht race to the New York newspapers in 1899.

Exciting as all this was, however, it wasn't until December 1901 that Marconi performed the feat that convinced the world radiotelegraphy was an extremely important discovery. A radio station was constructed at Poldhu, Cornwall, consisting of twenty masts, each over 200 feet high. Marconi then traveled to Newfoundland and, by using a kite to give the aerial wire greater height, he spent two days receiving radio signals from Cornwall, a distance of almost 200 miles. The world went wild.

Marconi had succeeded in connecting Europe with North America by sound, and everyone realized that it was simply a matter of time before similar and more advanced communication systems would connect every corner of the globe.

Marconi's involvement with radiotelegraphy continued throughout his lifetime. He was amply rewarded financially, at one time receiving an annual income of £20,000 from the British admiralty for the use of his radio equipment in the Navy. In 1924, following several years of work on a system of short-wave wireless communication, Marconi's company obtained a contract for the establishment of short-wave communication between Britain and its Commonwealth. He was also rewarded

GUGLIELMO MARCONI

Operating-room of the Western Union Company, New York.

85

with many honors and honorary degrees, including an Italian knighthood in 1902 and the Pulitzer Prize for physics in 1900.

When Marconi died in Rome in 1937, the Italian government accorded him a state funeral and he was buried in his home town of Bologna. Unlike many inventors who die broke or unappreciated, Marconi was a man who made good and lived to enjoy the results of his labor.

A woman listens to the radio in the comfort of her own home. The set pictured above is a 1923 model.

The great telegraphing room at the new offices of the Electric and International Telegraph Company, Bell Alley, Moorgate Street.

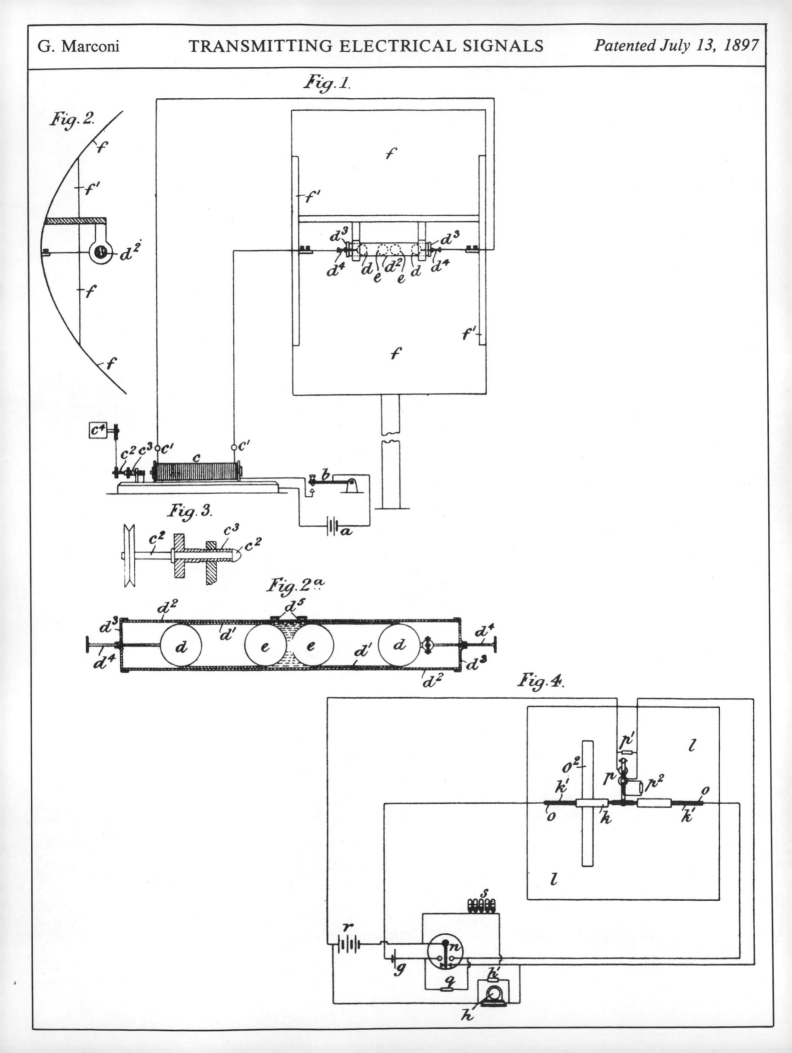

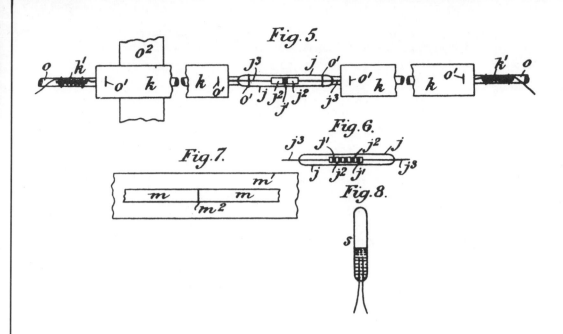

Fig. 5.

Fig. 6.

Fig. 7.

Fig. 8.

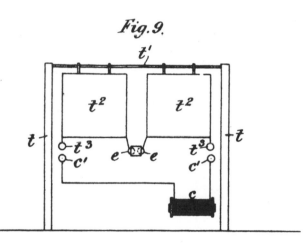

Fig. 9.

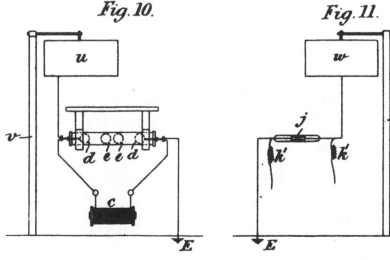

Fig. 10.

Fig. 11.

Marconi's transmitter and receiver. The concave reflectors helped to transmit the radio waves in the desired direction.

THOMAS ALVA EDISON

PHONOGRAPH

The majority of inventors hit upon a single good idea, early in their careers, and spend the rest of their lives trying to perfect it. This was not so in the case of Thomas Alva Edison. In his lifetime, he secured patents for 1,033 inventions, and his professional career spanned more than half a century.

Edison received his first patent for an electrical vote recorder, in 1868, at the age of twenty-one. A year later he took out a patent for an improved stock-ticker, which had initially been intended as a form of a typewriter; he had been called in by the executives of the Automatic Telegraph Company in New York to improve their automatic telegraph. While there, Edison was asked to assess the merits of a type-writing machine that had been brought to the company by a Mr. James Densmore and the machine's inventor, Mr. Christopher Latham Sholes. After weeks of examination, Edison advised the Telegraph Company to turn the invention down; he himself would be able to construct a much better machine for the same price. What actually resulted was not the typewriter Edison as the Automatic Telegraph Company had expected, but the stock-ticker.

Edison was essentially a self-taught man. His formal education consisted of three months in the public school of Port Huron, Michigan, where he spent most of his childhood. Unmotivated in school, and totally inept at mathematics, Edison nevertheless loved to read and by the time he was ten displayed a distinct taste for chemistry. When he was twelve, he was operating a successful newspaper and candy concession on railroad trains. By fifteen, he was employed full time as a telegraph operator, and in 1868 he went to work for Western Union in Boston, Massachusetts. His fascination with chemistry continued through this period and he was particularly interested in the applications of chemistry to telegraphy.

In 1869 he accepted a position with the Laws Gold Indicator Company in New York and formed a partnership with two other men as "electrical engineer" consultants. When the partnership was bought out in 1870, the proceeds from the sale enabled Edison to set up his own shop and devote his energies completely to his experiments. He devised duplex, quadruplex, and automatic telegraph systems, an electric pen that was the forerunner of the modern office duplicating machine, and a carbon transmitter, exhibited in 1877–78, that proved instrumental in advancing the science of telephony and

Edison developed a more sophisticated version of Bell's telephone transmitter. His device utilized sound waves in the mouthpiece (E) that vibrated against a carbon button (C) and changed the button's electrical resistance.

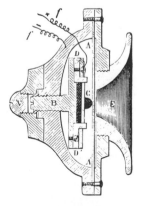

was of marked importance in adapting the Bell telephone for practical use.

In 1876 Edison moved his workshop and home to Menlo Park, New Jersey, and a year later applied for a patent on his "phonograph or speaking machine." Most of the work done on the original phonograph was performed by one of Edison's assistants, John Kruesi. Edison got the idea for recording the human voice while he was working on a device that would record Morse code messages. Kruesi took Edison's sketches for a voice recorder to the workshop and is reported to have been as surprised as anyone when the machine he had labored over repeated the words Edison had fed into it: "Mary had a little lamb."

Most inventions are improvements on the efforts of others; not so with the phonograph. The United States Patent Office could not discover one single patent or application for a patent that even remotely resembled Edison's machine. This machine, which originally sold for $18, recorded the voice on a cylinder that was covered with tin foil and turned manually with a crank. Edison continued his work on the machine and, in 1886, came out with a motor-driven version that used cylindrical wax records. These had been invented the previous year by Chichester A. Bell and Charles S. Tainter. The new machine became an instant success.

Edison is often given credit for the invention of the incandescent lamp, but, as with so many other inventions, what counts is not so much whose idea it is as who is able to make the idea work. In the case of the incandescent lamp, Edison did not invent it. The first electric light was invented by Sir Humphrey Davy, an English scientist from Cornwall. This lamp, produced in 1810, was known as an arc lamp, but proved to be too bright for use in the home. Edison in the United States and Joseph Swan in England both had the idea for a lamp that used a carbon filament inside a glass dome — essentially, the incandescent lamp — at approximately the same time. Swan managed to patent his invention in 1878; Edison followed a year later.

Although Edison cannot be credited with the incandescent lamp's discovery, no one can deny his importance in its development. Before finally hitting upon a successful model, he had invested more than $40,000 on experimentation. His first success with the lamp came when he was able to make a glass dome within which a loop of carbonized cotton thread glowed in a vacuum for more than forty hours. Further experimentation showed that tungsten was the best material to be used for the filament, and it is still used in most lamps today.

With the success of this discovery, Edison spent the next ten years of his life developing the use of electricity, making it accessible to the general public and finding viable methods for its generation. He not only made substantial improvements which enabled the lamp to be produced inexpensively and marketed universally, he was instrumental in establishing central stations that supplied these lamps with power.

In 1883, four years after an initial breakthrough with the incandescent lamp, Edison made his one major scientific discovery: he was able to show that the incandescent lamp could be used as a valve that would admit negative but not positive electricity. At the time he was unable to grasp the full implications of the discovery, but the "Edison effect" proved to be the basis for the vacuum tube, which is essential to modern radio.

THOMAS ALVA EDISON

Edison's phonograph recorded the voice on a cylinder that was cranked manually.

Illustration from Edison's patent application for an electric lamp.

One of Edison's chief assets was his ability to keep several ideas running through his mind simultaneously. While working on one invention he was able to think about and work on one or more others.

One technological advance he looked upon with great interest was the development of a machine that would make moving pictures. There had been many attempts in the early nineteenth century to perfect a system of cinematography. The English scientist Henry Fitton was one of the first people to come up with a viable idea on the subject. In 1826 he introduced what he called a "Thauma-tropical Amusement," which was essentially a small round box with several discs made out of cards inside it. Each disc had a design drawn on it. As the

discs were twirled about, these designs merged, giving the impression of a single moving image. In the years that followed, several more attempts were made to improve upon Fitton's device.

It wasn't until the 1870s, however, that photography was introduced into the situation. The English photographer Eadweard Muybridge tried to determine the positioning of a horse's legs in motion. He set up several cameras at equal intervals along the side of a race-track and attached a string to them which would trigger the cameras' shutters as the horse galloped past. Placing the pictures in a zoetrope for viewing, Muybridge discovered many things that were not apparent to the naked eye. The zoetrope, invented in France in 1860, consisted of a drum with vertical slots regularly placed around the sides. The

pictures were placed side by side within the drum. As the drum revolved, the viewer peered through different slots, deriving an impression of movement.

Several inventors pursued the moving picture goal simultaneously. In 1882 a French inventor exhibited a photographic gun which he had devised to take pictures of birds in flight. The same man invented a camera capable of taking sixty pictures a second. During this period many patents were taken out by French inventors for cameras and projectors.

Employing the discoveries of these other inventors, Edison worked on a moving picture and camera of his own. In his camera, which he called a kineto-graph, he used a flexible film that George Eastman, an American manufacturer, had invented in 1888. Edison called the

A family receiving a message from America.

93

projector a kinetoscope. The kinetoscope was a box almost four feet high. There was a small peephole in its side through which the viewer peered to see the pictures. Edison's original kinetoscope displayed fifty feet of film which ran for almost thirteen seconds. He was later able to slow the film down so that it ran for almost forty seconds.

Always wise to the market, Edison established several "kinetoscope parlors" where people could place a coin in a slot and view the films. He received a patent for his kinetoscope in 1891. Four years later, a man named Thomas Armat invented a machine that projected the image from the film onto a screen.

Other men had hit upon similar discoveries at the same time: Woodville Latham, an American; the Lumiere brothers from France; and Robert W. Paul of England. In 1896 Edison acquired the patent to Armat's machine and continued to make improvements on it. His idea was to coordinate the kinetoscope with his earlier invention, the phonograph. Unfortunately, every attempt he made to do so proved unsuccessful.

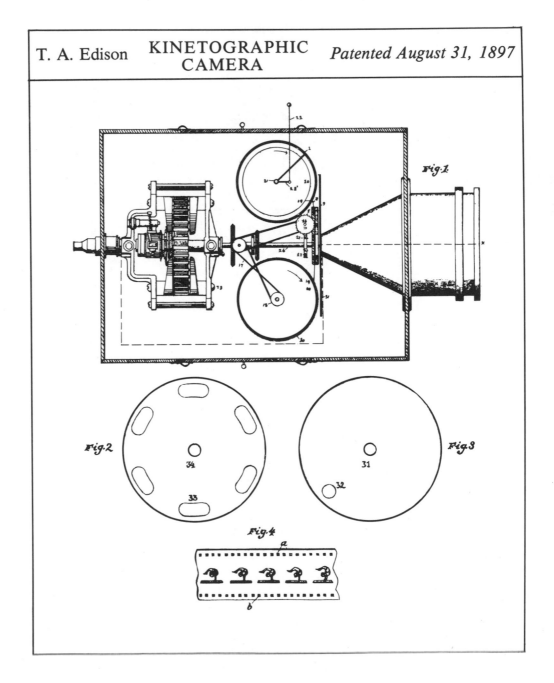

T. A. Edison   KINETOGRAPHIC CAMERA   *Patented August 31, 1897*

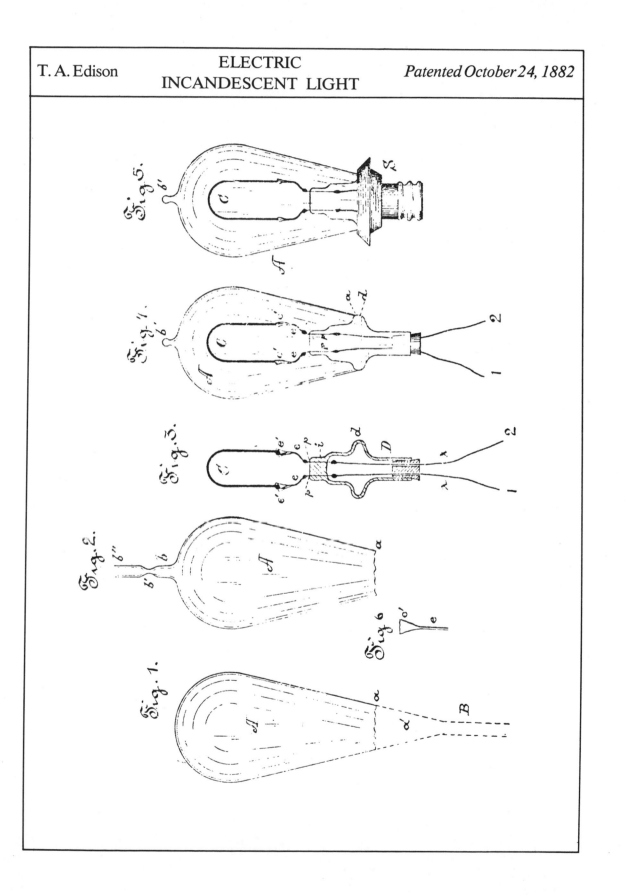

RUDOLF DIESEL

INTERNAL COMBUSTION ENGINE

Rudolph Diesel did not construct the first compression-ignition oil engine in existence, although the invention bears his name. In 1890 a Briton named Akroyd Stuart took out a patent for an oil engine that worked at low pressure, and a small locomotive with an engine based on Stuart's designs was made by the British firm of Hornsby and Sons in 1896. Diesel didn't take out his patents until 1894, and while he had been experimenting with engine designs since the 1880s, the first engine built according to his specifications was not constructed until 1897.

Born in 1858 in Paris, where he spent his childhood, Diesel studied engineering at Augsburg and Munich. He learned of the experiments of N.L.S. Carnot, a French scientist who was famous for a paper, published in 1824, entitled "Reflections on the Motive Power of Heat." In it, Carnot wrote that air, compressed to one-fourteenth of its original volume, might be used to ignite fuel. Diesel immediately began a series

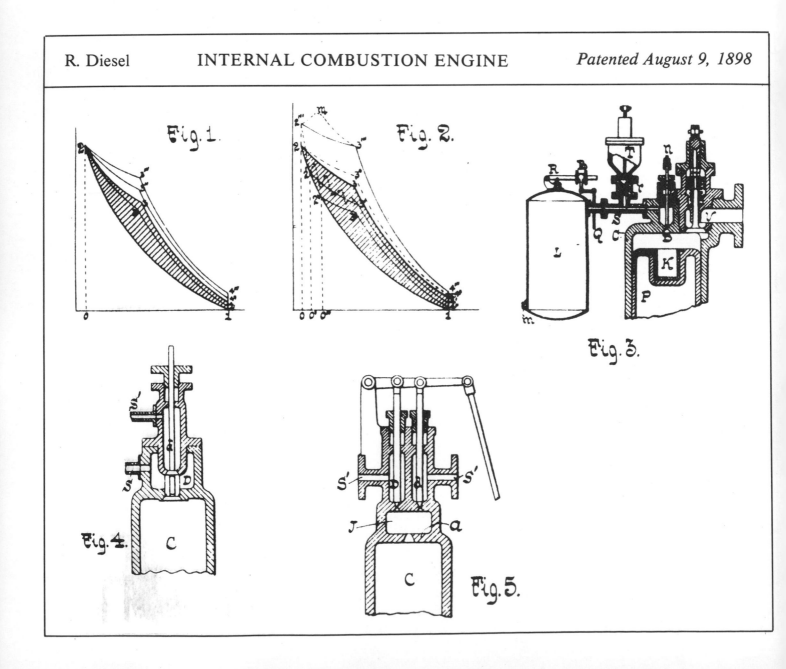

R. Diesel INTERNAL COMBUSTION ENGINE *Patented August 9, 1898*

Fig. 1.

Fig. 2.

Fig. 3.

Fig. 4.

Fig. 5.

of experiments based on Carnot's principles, as well as on those of the French engineer Alphonse Beau de Rochas, who, in 1862, had conceived of the fundamental principles upon which the modern automobile engine is based. Diesel's investigations led to the theory that, by compressing the air in the cylinder of an internal-combustion engine more than was considered standard at the time, he would achieve a higher level of fuel efficiency. In addition, the increased heat generated would spontaneously ignite the charge admitted into the cylinder, thereby eliminating the necessity of external fuel-ignition equipment.

Diesel worked for years before he came up with his "Working Processes for Internal-Combustion Engines," for which he was awarded a U.S. patent in February 1892. The following year he received another patent for improvements upon the first invention. Although these patents were not for diesel engines, they were instrumental in paving the way for that discovery, which Diesel actually patented in 1898.

RUDOLF DIESEL

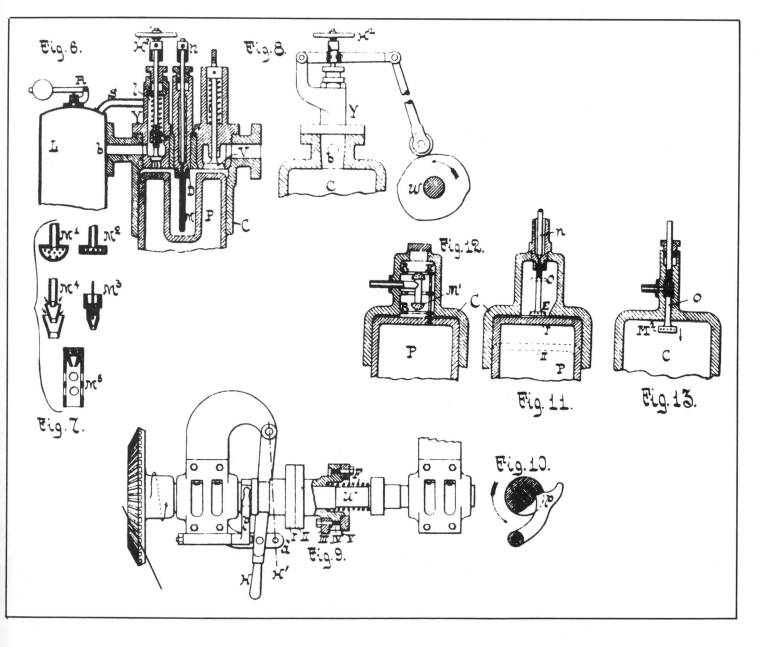

HENRY FORD

AUTOMOBILE

Henry Ford did not invent the automobile. Men such as Carl Benz and Gottlieb Daimler in Germany had been building gasoline-fueled vehicles since the 1880s. Their first such vehicles were three-wheeled machines, but by 1886 Daimler had come out with a four-wheeled automobile powered by a one-cylinder engine; Benz followed with his own version in 1890.

The first American to be granted a patent for an automobile was George B. Selden, who applied for one on May 8, 1879, but didn't build his machine until sixteen years later. Two other Americans, Charles E. and J. Frank Duryea, built a gasoline-powered automobile of their own in 1893. The idea for the car was Charles Duryea's; the actual manufacturing of it was done by Frank Duryea. The Duryea car was powered by a one-cylinder gasoline engine, and had an electric ignition. Its premiere run was made on September 21, 1893; two years later, in the first automobile race in America, J. Frank Duryea beat out all contenders over a distance of 54.36 miles. The Duryea traversed the distance roundtrip between Chicago and Evanston, Illinois, in 7 hours and 53 minutes.

Several other men were building motor vehicles at the same time. John William Lambert built one in 1890 and exhibited it in Ohio City, Ohio in 1891. Gottfried Schloemer of Milwaukee built an automobile in 1890, which still exists. Henry Nadig built an automobile that relied upon a wooden wick to fuel the cylinder by capillary attraction. Elwood G. Haynes came out with a gasoline-powered machine in 1894. Ransom E. Olds, of Reo and Oldsmobile fame, started research on the gasoline engine in the 1890s, as did Alexander Winton and James Ward Packard. Although there were more than fifty companies producing automobiles by 1898, the first commercially successful American-

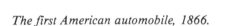

The first American automobile, 1866.

The automobile has evolved quite a bit since inventors began designing a machine that would run mechanically. In 1876, a man named Mathewson designed this steam horse for street railways.

made vehicle was a three-horsepower Oldsmobile. In 1901 the company sold 425 of them; in 1904, it sold 5,000.

Given the large number of competitive automobile manufacturers, naturally there were disputes regarding patents. Henry Ford, for one, contested the validity of Selden's patent. After an eight-year battle, the U.S. Circuit Court of Appeals upheld Selden's claim, but only for automobiles operating with a two-cycle engine. The Association of Licensed Automobile Manufacturers was established with the idea of promoting stability within the industry. As a consequence, an agreement was reached in

1915 which lasted till 1956, permitting the cross-licensing of patents.

In June 1903 Henry Ford established the Ford Motor Company. He sold his first car on July 23, and produced 1,700 automobiles within the first year. In 1909 Ford patented his Model T machine, having previously manufactured eight different models, bearing the letters A, B, C, F, K, N, R, and S. It was Ford's intention to take the automobile out of the realm of the luxury item and turn it into a universal necessity, which he accomplished within twenty years. By employing the techniques of mass production, an old idea that had never been

HENRY FORD

A new steam carriage (1884).

One interesting achievement combined a tricycle and a printing press. This particular model was invented in 1895.

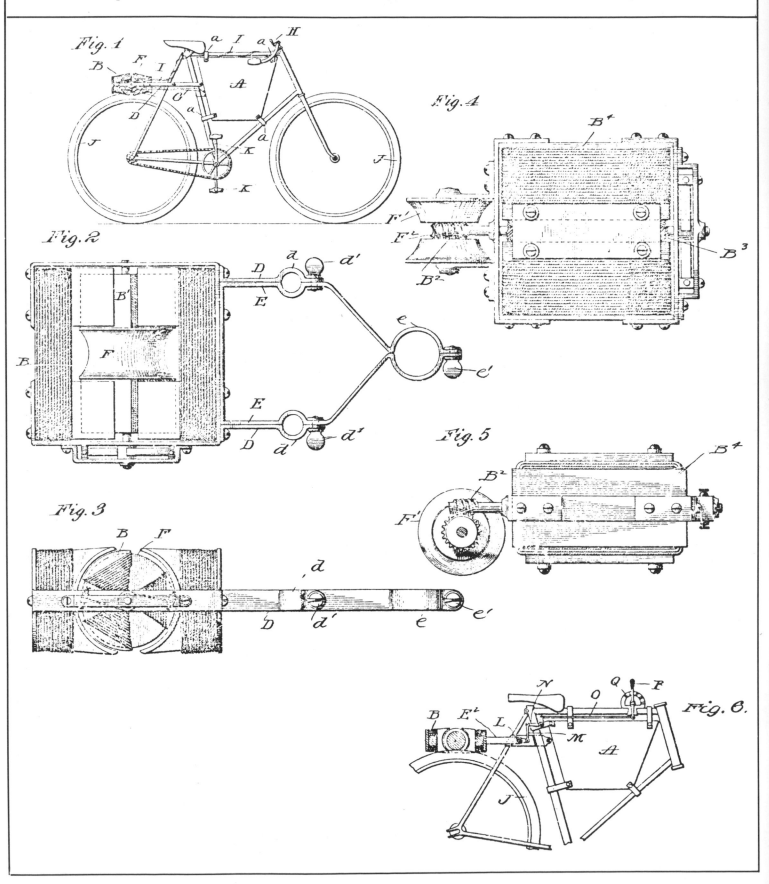

Fig. 1

Fig. 2

Fig. 3

Fig. 4

Fig. 5

Fig. 6.

fully realized, Ford built a financial empire that began, in 1903, with $100,000 capital and emerged, in 1927, (the year that the last of more than 15 million Model T's was made) with a surplus balance of almost $700 million.

Ford's business formula consisted of reducing the price of the product, increasing its sales, improving efficiency in production, increasing the output, and continuing the cycle. Obviously, his plan worked. He implemented the concept of a minimum wage, guaranteeing his workers at least $5 a day as compared to the standard salary in most U.S. manufacturing industries of $11 a week. In addition, Ford instituted a profit-sharing plan for his employees.

History shows Henry Ford to have been a man of extreme contradictions. He donated large sums of money to endow a modern hospital, yet subsidized a newspaper with an anti-Jewish slant. During World War I, with his company one of the most active contributors to the war machine, he chartered a ship

that sailed with a group of pacifists whose aim was to appeal to the various heads of state to "get the boys out of the trenches by Christmas." Although recognized as a leader of the new industrial order, Ford distrusted bankers and stockbrokers; he reinvested all of his profits back into the company, thereby

The Duryea Motor Wagon of 1896, complete with America's first pneumatic tires. The original Duryea car, manufactured in 1893, was powered by a one-cylinder engine and had an electric ignition system.

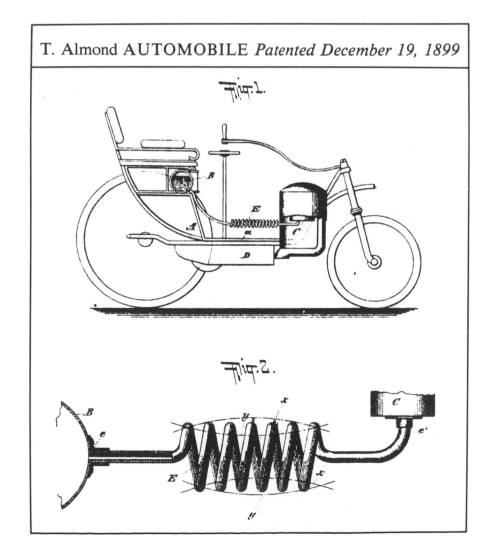

T. Almond AUTOMOBILE *Patented December 19, 1899*

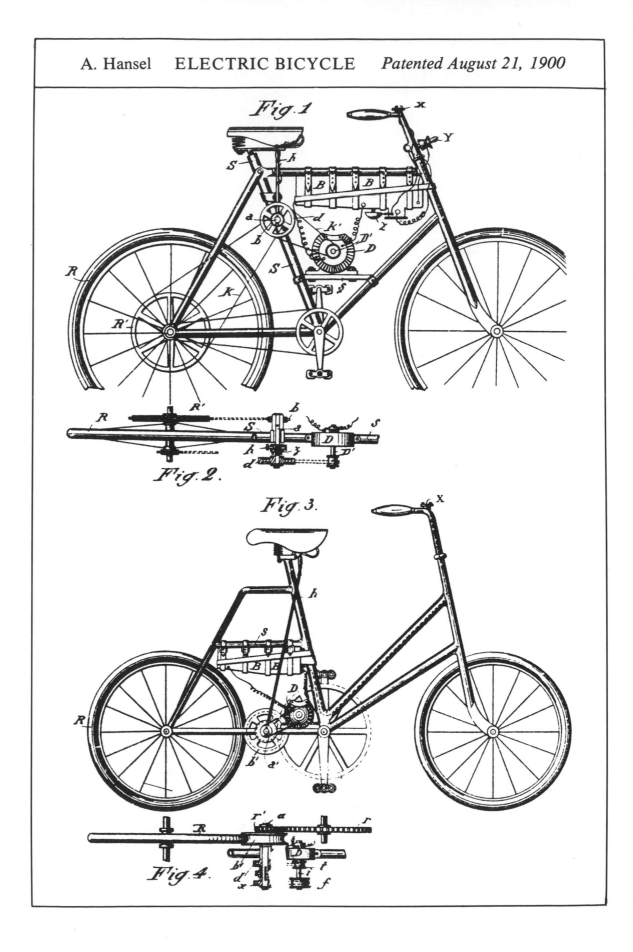

Fig. 1.

Fig. 2.

Fig. 3.

Fig. 4.

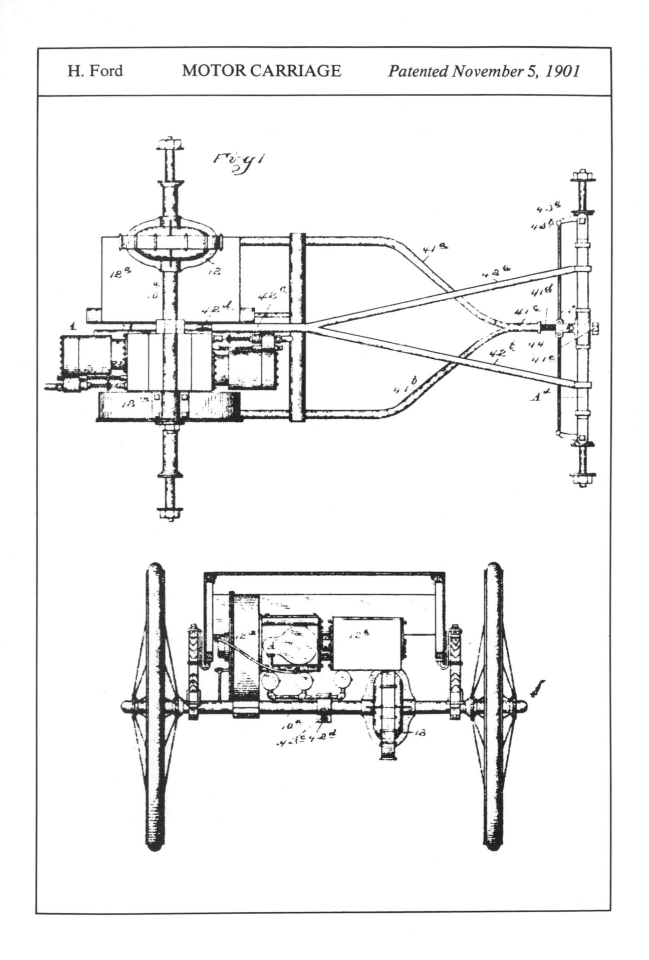

expanding through his earnings rather than by selling stock. Over the years, he gained complete control of the company by buying up most of its stock, and he ran things with an iron fist. Under his reign, the Ford Motor Company became the only company of comparable size to be controlled by one individual.

Ford, more than anyone, is responsible for the transformation of the world automotive industry. His use of the assembly line in automobile production revolutionized the industry by effecting a switch from the manufacture of customized cars for the wealthy few to the mass production of cars for the public at large. Following Ford's example, manufacturers in the United States and throughout Europe mass-produced automobiles for an eager world market.

In 1927, finally realizing that his Model T was outmoded, Ford completely reworked his original idea and came out with an entirely different car, the Model A. Five years later he manufactured an automobile with a V-8 engine, but his timing had been slow, and the premiere position of the Ford Motor Company had already been taken over by General Motors.

Dog cart powered by electricity (1888).

Electrical carriage (1888).

Henry Ford's classic Model T. The first Model T was built in 1908. By 1927, more than 15,000,000 had been produced.

Assembly line production.

The man who started it all, the late Henry Ford, shown with his first car, the 1896 Quadricycle, and the 10 millionth Ford car: a 1924 Model T.

1909 Ford Model T

1928 Model A Roadster

*Ford started something new in 1929 with the auto industry's
first mass-produced station wagon.*

1929 Ford Model A

1932 Ford V-8

1979 Ford Mustang prototype

KING CAMP GILLETTE

RAZOR

Shaving one's facial hair is by no means a modern idea. Archaeologists have discovered cave paintings which depict men shaving with sharpened flints and pieces of shell that date back over 20,000 years. The ancient Greeks shaved their faces, much to the scorn of the Romans, who finally followed suit about 2,400 years ago. (A shaven face was much easier to keep clean in an age when personal hygiene was difficult to maintain.)

The early, crude shaving utensils evolved into more sophisticated ones. Flints and shells gave way to bronze, iron, and, occasionally, even gold razors. Through the centuries there were advances in design, culminating in the late eighteenth century with the invention of the first safety razor. In 1762, in France, Jean-Jacques Perret invented a safety razor that utilized a metal guard, placed along one edge of the blade, to prevent the blade from slicing into the shaver's skin. Almost seventy years later, in Sheffield, England, a similar but improved design, utilizing a steel blade, was introduced. And in 1880, an American designed a fixed-blade razor in the shape of a T, the pattern that is still used today.

In 1895, while using a T-shaped fixed-blade razor, King Camp Gillette, an American salesman, thought of a way to make a lot of money. For years he had been wracking his brain trying to invent a product that would make him rich. The fame and fortune syndrome

Sheffield razors were first produced in 1830 in Sheffield, England, by a manufacturer named W. Fenney.

was definitely what motivated him. Now he had finally discovered it. The major problem with the razor he was using was that the blade never stayed sharp for very long. After a few shaves, it either had to be resharpened or thrown away. Gillette realized that razors were too expensive to be disposed of whenever the blade became dull. Why not construct a disposable blade, he reasoned, so that the user could keep his razor but insert a fresh, sharp blade whenever the old one became dull?

Unfortunately, Gillette was not a skilled metalworker. His talents were in the field of sales. For six years he struggled to develop a blade that would be thin enough to be manufactured cheaply and yet strong enough to take a sharp edge. Then a friend introduced him to an engineer named William Nickerson who took his idea and turned it into reality.

Gillette quickly patented the invention in 1904 and set his sights on developing the market. He had no money of his own, so he tried to get financial backing. He manufactured several razors at his own expense and gave them away. A wealthy friend soon saw the money-making possibilities of Gillette's invention and put up the money to produce the razors on a mass-market scale. From an initial sale of 51 razors and 168 blades during Gillette's first year in business, the company grew so rapidly that in 1908 it produced 300,000 razors and 14 million blades. Gillette's quest for fame and fortune succeeded; not only did he become a fabulously wealthy man, but his name is still a household word.

KING CAMP GILLETTE

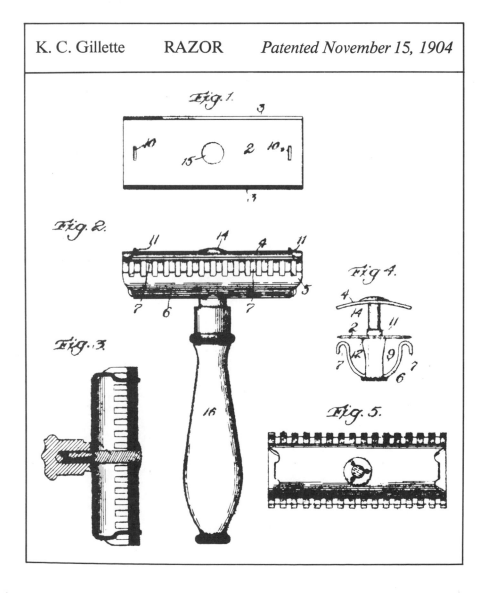

K. C. Gillette   RAZOR   *Patented November 15, 1904*

A pre-World War I advertisement that appeared in Great Britain. Ads such as this increased Gillette's sales from 51 razors and 168 blades in 1904 to 300,000 razors and 14,000,000 blades in 1908.

The Gillette safety razor and disposable blades revolutionized the cosmetics industry in the U.S.

112

King Camp Gillette as he appeared on an early razor package.

WRIGHT BROTHERS

AIRPLANE

SUCCESS FOUR FLIGHTS THURSDAY
MORNING ALL AGAINST TWENTY
ONE MILE WIND STARTED FROM
LEVEL WITH ENGINE POWER ALONE
AVERAGE SPEED THROUGH AIR
THIRTY ONE MILES LONGEST 57
SECONDS INFORM PRESS HOME
CHRISTMAS

ORVILLE WRIGHT

Man had learned to fly. On the morning of December 17, 1903, Orville Wright was the first man in history to make a fully powered, totally controlled, and sustained flight through the air. His very first flight lasted all of twelve seconds and covered a distance of some 120 feet down a beach before the machine landed safely in the sand. Three subsequent flights that same day, with Orville and his brother Wilbur taking turns as pilot, proved even more successful; the longest flight covered a distance of 852 feet and, as Orville's telegram to his father relates, lasted 57 seconds. What civilization had attempted for centuries, ever since Icarus made his vertiginous fall, was finally accomplished on Kill Devil Hill at Kitty Hawk, North Carolina.

Back home in Dayton, Wilbur and Orville Wright operated the Wright Cycle Co., which they formed in 1892. They sold and repaired bicycles, and received moderate acclaim for their manufacture of the "Van Cleve" bicycle.

Interested in aviation, they eagerly followed the exploits of Gustav Otto Lilienthal, the German aeronaut, who had received worldwide attention for his sensational experiments with gliders. Lilienthal had observed the flight of birds and written a book about it that was published in 1889. He employed this information in the making and flying of several different types of gliders, some with one wing, some with two. Much in the same way that a hang glider operates today, Lilienthal fitted himself into his machines and hung from them, using his hands and arms to hold on. His only means of altering the course of the glider in midflight was by swinging his legs or tilting his body. One of his most successful flights carried him a distance of 750 feet. Unfortunately, his glider was extremely sensitive to changes in wind direction, and one day Lilienthal was killed when an unexpected gust of wind caused it to crash.

Other men such as Octave Chanute and S.P. Langley conducted experiments and research with modes of aviation. Langley received the sanction and financial backing of the United States Army for his construction of an aerodrome, which met an untimely end when it failed to clear its launching pad and unceremoniously plunged into the waters of the Potomac.

Man's first attempts to fly were often quite imaginative.

Whereas Langley focused on the theory of flight, the Wright brothers concerned themselves with the reality of flying. Influenced by Lilienthal, they first experimented with gliders in order to understand the patterns of air currents and wind flow. Their first and major discovery occurred in 1899, when they realized that birds maintain their balance in flight by a simple adjustment of the tips of their wings. After several attempts at producing this in their gliders, they finally hit upon a way of warping the wings to achieve the same end. As Orville remarked: "The basic idea was the adjustment of the wings to the right and left sides to different angles so as to secure different lifts on the opposite wings." This device, which is now known as aileron control, served as the model upon which all further developments of lateral control were based.

The brothers conducted several experiments in their native Ohio before traveling to Kitty Hawk which was recommended by the United States Weather Bureau as ideally suited to their need to conduct more exhaustive research. Their first experiments were conducted in October 1901. Manned, the glider stayed aloft for two minutes; flown as a kite, with a man lying prone on the lower wing, the glider stayed up for ten minutes. Although the results were not as successful as they had anticipated, the experiment proved the effectiveness of their warping system, and the brothers returned to Dayton to make further improvements and to design a machine that could be propelled by an internal-combustion engine.

Orville designed a movable rudder which replaced the immovable rear fins, and Wilbur connected them to the hip-operated cradle that warped the wings. After several months of work and many atttempts to interest the automotive industry in building a lightweight motor, the Wrights, with the assistance of their mechanic, Charlie Taylor, built a motor that weighed less than 180 pounds. They used their own wind tunnel to develop adequate propellors and built a completely new "glider" to accommodate all of these changes.

In late September 1903, they again traveled to Kitty Hawk. They shipped their machine unassembled and spent almost a month reassembling it on loca-

tion. After several delays caused by damage to the machine and by weather problems, the Wright Brothers' *Flyer* was ready for testing on December 14. Wilbur won the honor of making the first flight, but the machine stalled as it was gaining altitude and crashed after only three and a half seconds of flight. Once minor repairs were completed, the brothers tried it again. This time Orville had the honor, and the flight was a success.

Despite the triumph of these flights, the United States was unable to see the importance of the airplane. The Wrights began a worldwide marketing campaign, during which, in 1908, Wilbur traveled to Europe to sell French rights. He flew at Le Mans and Pau in France, as well as in Rome, and received international attention. Simultaneously, Orville conducted a series of tests at Fort Myer, Virginia, where he made 57 complete circles at an altitude of 120 feet and remained aloft for 62 minutes. The formation of the American Wright Company shortly thereafter brought about the quick sale of airplane rights in England, France, Italy, and the United States.

Although Wilbur died of typhoid fever in 1912, Orville devoted himself almost exclusively to further research until his own death in 1948.

WILBUR WRIGHT

ORVILLE WRIGHT

J. A. C. Charles, a Frenchman, designed the first hydrogen balloon. Charles's balloon, called a Charliere, rose from the Tuileries Gardens in Paris on December 1, 1783. It wasn't long before balloons of all shapes and sizes rose above Europe and America.

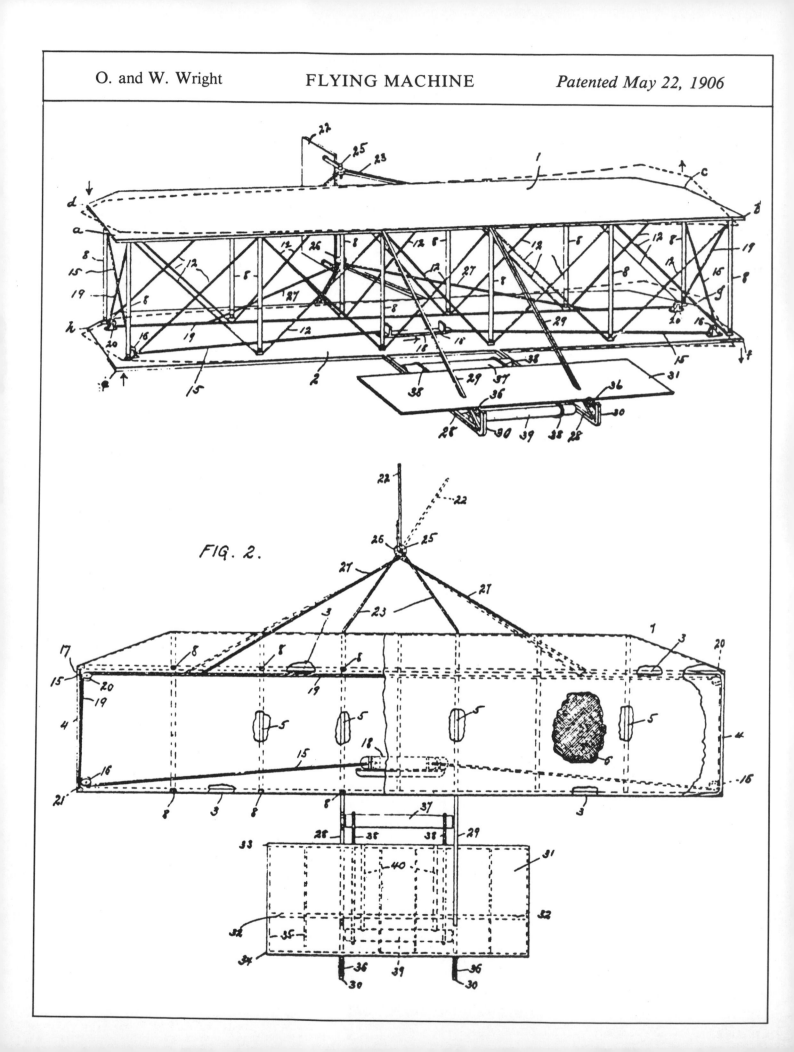

FIG. 2.

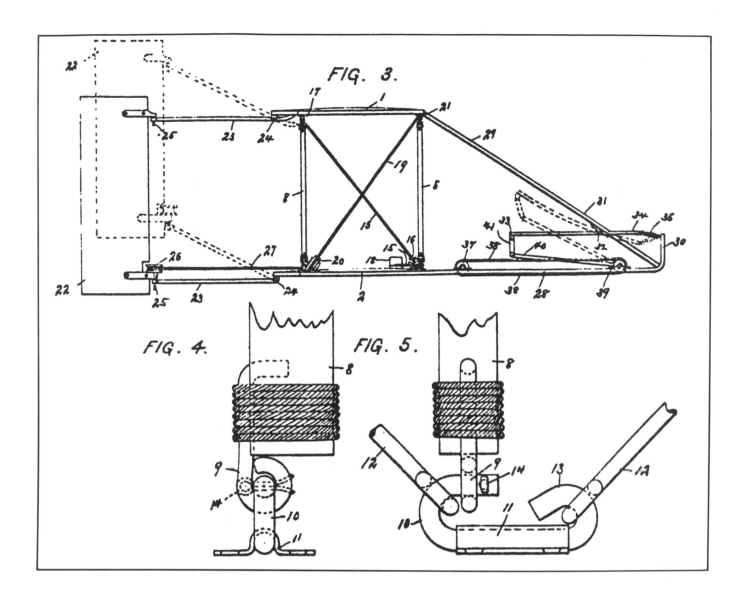

The Wright Brothers's first plane, being test run at Kitty Hawk, North Carolina on December 17, 1903.

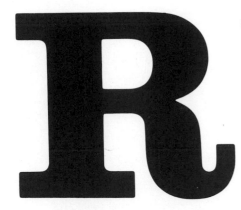

ROCKET

ROBERT H. GODDARD

On April 12, 1961, Yuri Gagarin, a Soviet cosmonaut, became the first man in history to orbit the earth in a rocket. The United States quickly followed suit, and Alan Shepard, an astronaut from New Hampshire, became the first American to be launched into space. Although mankind had indulged in rocketry and space flight for centuries, the first truly scientific experiments with rockets were not conducted until the twentieth century, when Robert Hutchings Goddard, a student at Worcester Polytechnic Institute in Massachusetts, began to design rockets propelled by solid fuel. Goddard, generally considered the father of U.S. rocketry, was graduated from the Institute and went on to Clark University where he studied for his doctorate. In 1911, upon graduation from Clark, he was hired as a member of the Clark faculty. Throughout all this time, he was obsessed with the idea of penetrating the outer reaches of space by means of rockets.

In 1919 he published the results of his research, which had been funded by the Smithsonian Institution, in a report entitled "A Method of Researching Extreme Altitudes." At the time he wrote the article, Goddard was primarily using solid fuel to propel his devices. Solid fuel consisted of mixtures of resin, rubber, or asphalt with an oxygen compound. He later switched to liquid fuel, which utilizes liquid oxygen that produces the combustion of alcohol, peroxide, and aniline or nitric acid. Goddard built his first liquid-fueled rocket in 1926; four years later he launched a rocket that soared 2,000 feet in the air at a speed of 500 miles per hour.

Financed by grants from Clark University and the Guggenheim Foundation, Goddard traveled to New Mexico, where he conducted a series of experiments with liquid fuel rockets and gyroscope controls at a desert site outside of Rosewall. These experiments showed that a vacuum provides better conditions for rocket operation than the atmosphere. Out of this research, Goddard designed a system of space travel whereby rockets could journey through space in a series of stages as a means of reaching the moon. As in the case with most inventors, however, public response to Goddard's theories was skeptical. It took time to develop his theories to the point of practicability, and the American public looked upon him as somewhat "moon mad."

But Goddard continued to experiment, and during World War II he was employed by the United States Navy to design rocket motors and jet-assisted takeoff devices (known as JATO) for aircraft. He moved his laboratory to Annapolis, Maryland and continued his work until his death on August 10, 1945. Goddard's patents, now owned by the Guggenheim Foundation, were used by the Nazis to develop their own V-2 rocket program. Under the leadership of Hermann Oberth, a team of German scientists had been heavily involved in space research throughout the 1920s and 1930s. After the war, when most of the Reich's top space

Dr. Robert H. Goddard with the first liquid propellant rocket to be flown, March 16, 1926, at Auburn, Massachusetts.

engineers were imported to the United States to develop a U.S. space program, Goddard's patents and technological advances were utilized. The patent infringements initiated by the Germans during the war continued in this country as well, until the 1950s, when the Guggenheim Foundation was awarded an unprecedented $1 million settlement by the courts for the patent law violations committed by the U.S. government.

In posthumous recognition of Goddard's contribution to space technology, the National Aeronautics and Space Administration (NASA) dedicated its facility in Greenbelt, Maryland to him, naming it the Goddard Space Flight Center.

DR. ROBERT H. GODDARD

R. H. Goddard ROCKET APPARATUS *Patented July 7, 1914*

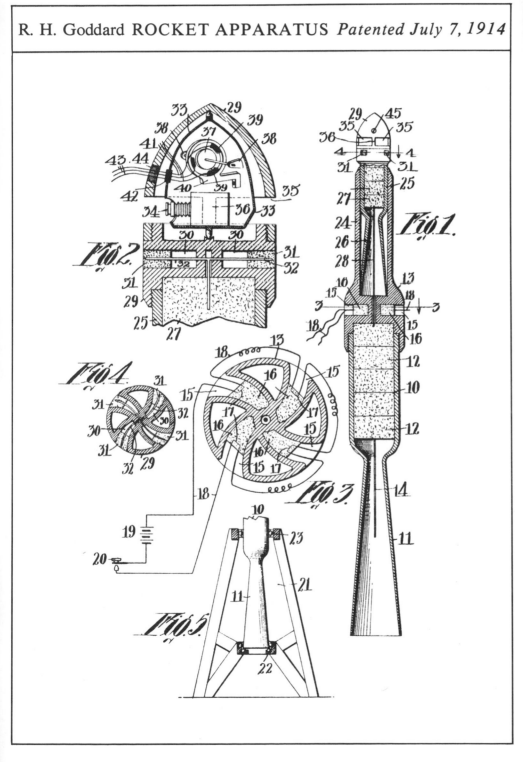

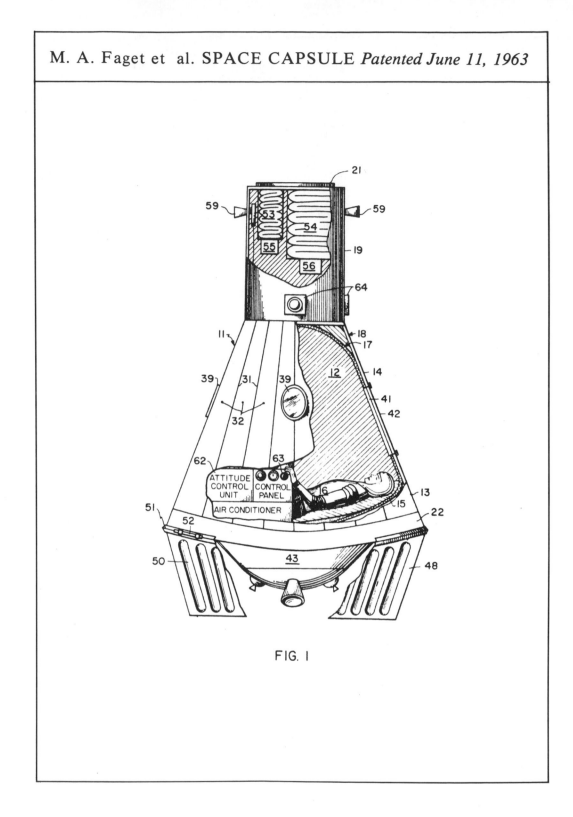

FIG. 1

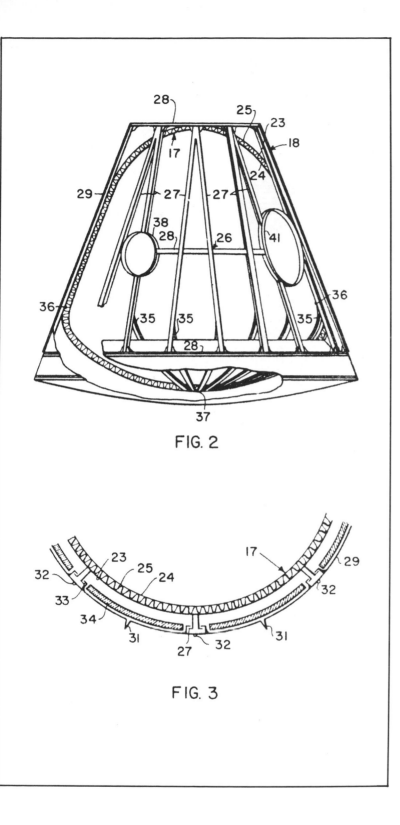

FIG. 2

FIG. 3

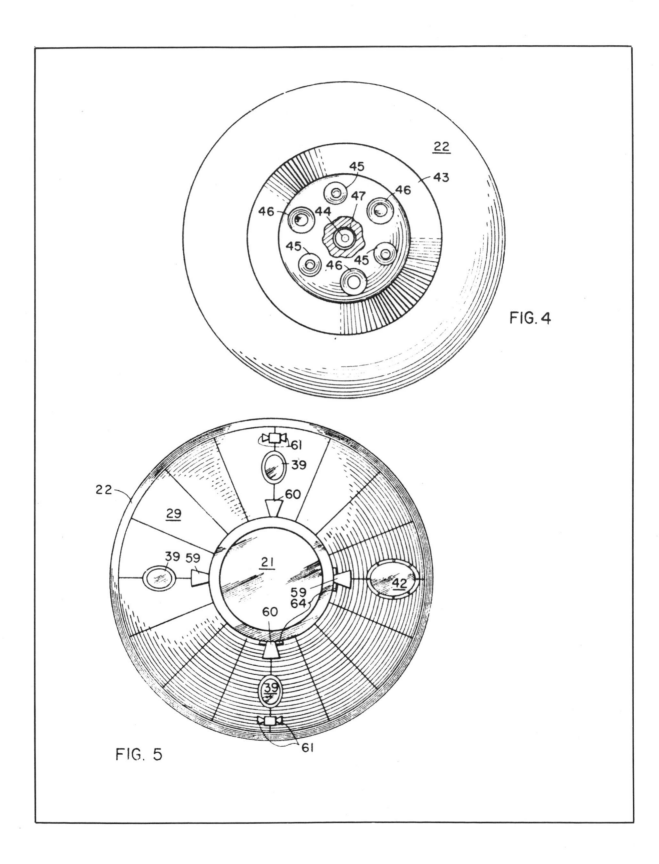

FIG. 4

FIG. 5

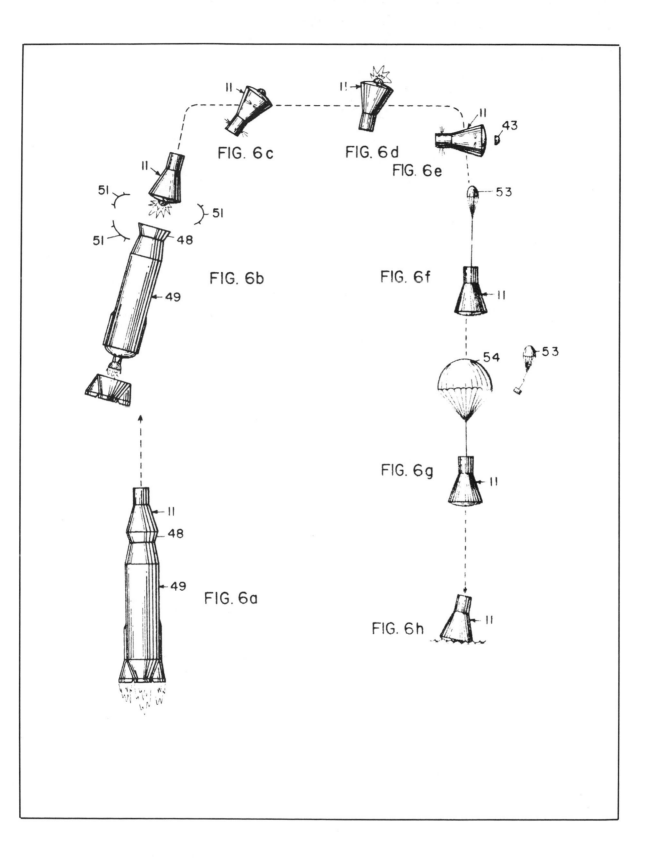

FIG. 6c

FIG. 6d

FIG. 6e

FIG. 6b

FIG. 6f

FIG. 6g

FIG. 6a

FIG. 6h

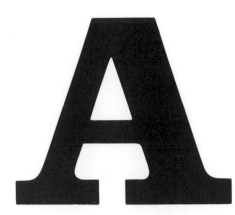

ARTHUR L. SCHAWLOW
C.H. TOWNES

LASER

In 1958 Arthur L. Schawlow and C.H. Townes, two research physicists, proposed the possibilities of a laser, which they described as a device for the amplification or generation of coherent light waves. Unlike waves that are emitted from an ordinary light bulb, the laser produces a beam of light that does not diffuse. It organizes the energy waves that are emitted by a stimulated atom in such a way that they move in the same direction, at the same frequency, and in complete accord with the stimulating radiation. The resultant wave band is similar to that of a radio frequency oscillator, but it utilizes the infrared and visible-light portions of the spectrum rather than comparatively low-frequency radio waves. Theoretically, this band is able to power 80 million television channels or 800 million simultaneous telephone conversations. Because the beam of a laser can be directed and focused to an extremely fine point, it has been used for such diverse purposes as radar tracking and range finding, welding, and surgery.

In 1960 T.H. Maiman produced the first working laser. Using a crystal of synthetic ruby as the active material, Maiman's laser required so powerful a light source that it could only operate for a short time. The ruby laser, which emitted a red light, is known as a pulsed laser and is used when a laser that can give an extremely high peak power is

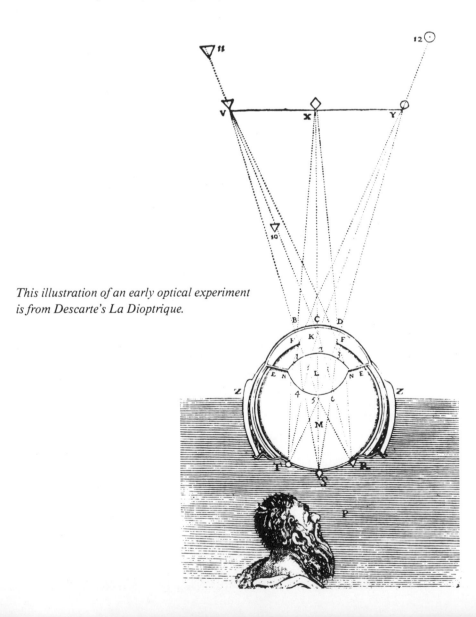

This illustration of an early optical experiment is from Descarte's La Dioptrique.

needed. Continuously operating lasers have since been invented which give very high frequency stability and purity. Since 1960, many different kinds of lasers have been developed, utilizing light, electrical discharges, or chemical processes for their energy. The possibilities for practical application seem limitless.

One application is in the field of photography. In 1947 a man named Dennis Gabor of the Imperial College of Science and Technology in London discovered a process of photography known as wave-front reconstruction. Unlike conventional photography that records a two-dimensional image of a three-dimensional object, wave-front reconstruction records the reflected light waves themselves. Although this photographic record, called a hologram, does not look anything like the original object, it contains all the information that a

two-dimensional photograph contains plus much more — all of it depicted on the hologram in the form of photographic code. Making sense out of the holographic code is the reconstruction process. *Scientific American*, in an article in its June 1965 issue, described the reconstruction process: "In this stage, the captured waves are in effect released from the hologram record, whereupon they proceed onward, oblivious to the time lapse in their history. The reconstructed waves are indistinguishable from the original waves and are in effect capable of all the phenomena that characterize the original waves. For example, they can be passed through a lens and brought to a focus, thereby forming an image of the original object — even though the original object has long since been removed! If the reconstructed waves are intercepted by the eye of an observer, the effect is exactly as if the

Metals being welded by laser for industrial purposes at Control Laser Corporation in Florida.

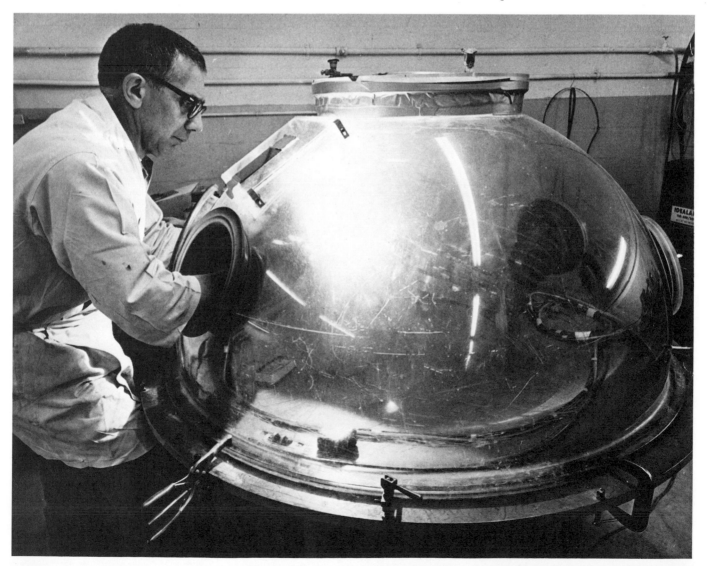

original waves had been observed: the observer sees what to all appearances is the original object itself in full three-dimensional form, complete with parallax (the apparent displacement of an object when seen from different directions) and many other effects that occur in the normal seeing process."

The invention of the laser greatly advanced the experimentation being carried on with holograms. Lasers provide the adequate source of coherent light that was needed to obtain high-quality three-dimensional images. Two scientists, Emmett N. Leith and Juris Upatnieks, who wrote the article for *Scientific American*, were the first to utilize lasers in holography in their laboratory at the University of Michigan in the early 1960s.

Although scientists are aware of the laser's potential usefulness in such areas as communication and medicine, the technology necessary for bringing it to fruition has not always been adequately developed. New processes and machinery are constantly being developed to advance this exciting field, and scientists and inventors are working overtime to find new ways to implement this exceptional energy source.

Lageos, a laser geodynamic satellite, was launched in April 1976. Lageos was designed to demonstrate and employ the capability of laser tracking techniques to make accurate determination of the earth's crustal and rotational motions.

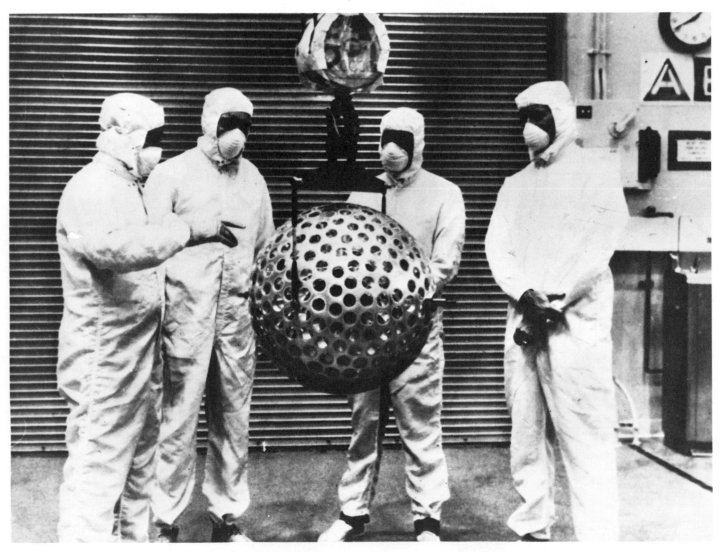

PHOTO CREDITS

The author and publisher acknowledge with thanks the use of photographs from the following sources:

Meredith Hooper, Everyday Inventions (Taplinger Publishing Co., 1976) ©Meredith Hooper 1972. Reprinted by permission. p. 52, 67

Drawings by Gardner Rea; ©1954. The New Yorker Magazine, Inc. p. 30

Cynthia A. Hoover, The History of Music Machines (Drake Publishers, 1975) p. 86

Cambridge University Press p. 15

Codex p. 13, 14, 15

Culver p. 43, 115, 117, 119, 121

Doubleday & Co. p. 19

Ford Motor Co. p. 99, 105, 106, 107, 108, 109

Gillette p. 11, 112, 113

Goldschmidt, E. P. & Co. Limited

The Grolier Society , p. 39, 43

Hutchinson & Co.

Illustrated Newspapers, Ltd. p. 31, 58

LaNature p. 68, 70

London News & Sketch, Ltd. p. 13, 16, 93, 12

Main Street Press p. 31

McGraw-Hill, New York p. 105

The National Historical Society, Gettysburg p.34, 35

Pach Brothers Photographers p. 69

Punch Magazine p. 13, 20, 32

Rocher, Monaco p. 18

Scientific American p. 69, 98-99, 104

Simon & Schuster p. 17, 90

Stage

U.P.I. p. 85, 91, 125, 126

Westinghouse Electric Co. p. 65